STEEL DETAILING IN CAD FORMAT

Kamel A. Zayat

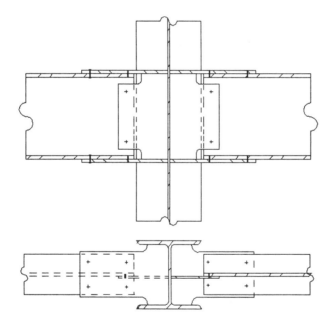

A WILEY-INTERSCIENCE PUBLICATION

JOHN WILEY & SONS, INC.

New York . Chichester . Brisbane . Toronto . Singapore

REQUIREMENTS:
An IBM PC family computer or compatible computer
with 256K minimum memory, a 3.5" high-density floppy drive,
PC DOS, MS DOS, or DR DOS Version 2.0 or later, and a printer.

Copyright© 1995 by John Wiley & Sons, Inc.

AutoCAD ® LT is a registered trademark of Autodesk, Inc.

This publication is designed to provide accurate and
authoritative information in regard to the subject
matter covered. It is sold with the understanding that
the publisher is not engaged in rendering legal, accounting,
or other professional services. If legal advice or other
expert assistance is required, the services of a competent
professional person should be sought.

Library of Congress Cataloging-in-Publication Data

Zayat, K. A. (Kamal A.), 1957-
    Steel detailing in CAD format / K.A. Zayat.
        p.   cm.
    Includes index.
    ISBN 0-471-10992-4 (cloth : acid-free paper)
    1. Building, Iron and steel—Details—Drawings. 2. Structural
engineering—Data processing. 3. Computer-aided design. I. Title.
TA684.Z39 1995
6241. 1'821'0285—dc20   94-41842

Printed in the United States of America

10 9 8 7 6 5 4 3 2 1

The winner is not the stronger,

the winner is the one who can stand longer.

K.A. Zayat

Preface

Steel Detailing in CAD Format is a reference library for <u>steel building construction</u>. Engineers, architects, and contractors will find the book useful for daily use and practice. It provides structural details for building, which is a major factor in building design.

Using a library of standard details saves much of the engineering time and money that was spent drawing the same details many times.

This steel detailing book covers the two most important forces: vertical and lateral forces. For each detail, the lateral resistance was carefully considered.

The details have been created using professional experience to maintain economical construction costs and to create safe buildings.

The book is divided into nineteen chapters - Base Plate, Lateral Brace, Bracket, Composite Section, Metal Deck, Foundation, Moment Connection, Opening, Pin Connection, Roof Details, Build-Up Section, Shear Connection, Slide Connection, Splice Connection, Stair Details, Metal Studs, Tension Connection, Torsion Connection, Trusses. The details in each chapter are further categorized by connections. For example, in the Moment Connections chapter there is a set of details for horizontal beams connected to columns, another set for sloped beams connected to columns. This method of dividing the chapters helps the user to easily become familiar with this reference book.

Some of the details are combinations of steel and wood, or steel and concrete.

We added some sizes and dimensions on some details, just to give an idea about the way to write dimensions and sizes on the details. All the sizes or dimensions in the book may not be used without structural calculation and confirmation with the official code.

We have provided each detail on an 8-1/2 x 11 sheet. Each sheet contains the drawing itself, a description of the detail, a check list which summarizes the detail, and a small plan view which suggests the location on the structural drawing where the detail can be applied.

Every detail has been drawn using CAD software. We believe many of our drawings are typical and can be copied directly from the book and attached to structural drawings. In addition, each drawing is saved in a separate file in .DXF format, which can be loaded into most CAD software programs to be modified as needed and plotted or printed. This system enables the user to create the needed details quickly.

The table of contents will give the user a good overview of the selection of details included, and provide an easy way to locate individual details. The detailed index is also useful for locating details which apply to particular situations. In any case, after using the book a few times, the user will be able to find the needed details very quickly.

Steel Detailing in CAD Format is the second of our series of books intended to computerize the drawing of details for any engineer. The ultimate objective of CASA utility is to computerize the drawing of details for any engineer. The ultimate objective of CASA Utility is to computerize as much as possible the tedious part of the work of engineering. In this way, the engineer will be able to create the structural calculation package and the detailing as quickly and as economically as possible.

We suggest that each user update his or her own library, starting with *Steel Detailing in CAD Format* and continuing with an organizational method such as we provide in our book.

I would like to thank all the people at John Wiley & Sons for giving me the opportunity to publish this book.

CONTENTS

CHAPTER 3: BRACKET

CHAPTER 4: COMPOSITE SECTION STEEL/CONCRETE

CHAPTER 5: METAL DECK

CHAPTER 6: FOUNDATION

CHAPTER 7: STEEL MOMENT CONNECTION

CHAPTER 8: OPENING THROUGH STRUCTURAL MEMBER

CHAPTER 9: PIN CONNECTION STEEL DETAILS

CHAPTER 10: ROOF STRUCTURAL STEEL DETAILS

CHAPTER 11: BUILD-UP STEEL SECTION

CHAPTER 13: CONNECTION WITH LONGITUDINAL SLIDING

Computer Requirements

The enclosed diskettes require an IBM-PC compatible computer capable of running DXF--capable CAD software with DOS version 2.0 or later. It can be used in both DOS and Windows environments. To load the files into your CAD software, consult your software manual.

The files on the diskettes have been saved in DXF format and compressed to save diskette space. The file names and page numbers are listed in the Table of Contents. The name of each file appears at the bottom right corner of each drawing in the book. The file names have been organized in the way in which they appear in the book. For example, the Base Plate detail file names are:

SBASE1.DXF
SBASE2.DXF
SBASE3.DXF
...
For the moment connections details, the file names are:

MOMENT1.DXF
MOMENT2.DXF
MOMENT3.DXF
...
For the foundation details, the file names are:

FND1.DXF
FND2.DXF
FND3.DXF

How to Make a Backup Diskette

Before you start to use the enclosed diskettes, we strongly recommend that you make a backup copies of the originals. Making a backup copy of the diskettes allows you to have a clean set of files saved in case you accidentally change or delete a file. Remember, however, that a backup diskette is for your own personal use only. Any other use of the backup diskettes violates copyright law. Please take the time now to make the backup copy, using the instructions below:

A.   If your computer has two floppy disk drives...

1. Insert your DOS disk #1 into drive A of your computer.
2. Insert a blank disk into drive B of your computer.
3. At the **A:>**, type **DISKCOPY A: B:** and press Enter. You will be prompted by DOS to place the source disk into drive A.
4. Place the <u>Steel Detailing</u> diskette into drive A. Follow the directions on the screen to complete the copy.
5. When you are through, remove the new backup diskette from drive B and label it immediately. Remove the original <u>Steel Detailing</u> disk from drive A and store it in a safe place.

Follow the instructions above to make a backup copy of disk #2.

B.   If your computer has one floppy disk drive and a hard drive...

If you have an internal hard drive on your computer, you can copy the files from the enclosed diskettes directly onto your hard disk drive, in lieu of making a backup copies, by following the installation instructions on the following pages.

Installing the Diskette

The enclosed diskettes contain over 200 individual files in a compressed format. In order to use the files, you must run the installation program for the diskettes .

You can install the diskettes onto your computer by following these steps:

1. Insert the Steel Detailing disk #1 into drive A of your computer. Type **A:\INSTALL** and press Enter.

2. The installation program will be loaded. After the title screen appears, you will be given the options shown in Figure 1.

3. The following Menu Selections will be listed: Edit Destination Paths, Select Destination Drive, Toggle Overwrite Mode, Select Groups to Install, and Start Installation.

4. The **Destination Path** is the name of the defaᵘlt directory to store the data files. The default directory name is STEEL. To change this name, press Enter, hit the letter **P**, type in the name of the directory you wish to use, and press Enter.

5. **Select Destination Drive** gives you the option of installing the files onto a hard disk drive C:\ or, if you wish, onto a different drive.

6. The **Toggle Overwrite Mode** pertains to the directories and files you already have on your hard drive. Do not give the default directory the same name as existing directories on your hard drive or the installation program will overwrite or delete any pre-existing directories of the same name. The safest option to protect existing data on your computer is OVERWRITE NEVER.

7. The **Select Groups to Install** option allows you to install each subdirectory from the diskettes onto your hard drive one by one. The subdirectories on the diskettes are named for the different chapters of the book. For example, subdirectory BRACKET contains all of the files that are covered under the chapter of the same name. If you wish to install the entire directory at once, tab down to Start Installation and press Enter. To install just one of the chapters, press Enter, hit the letter **G**, select the chapter(s) you do not want to install, and hit Enter again.

The files are now successfully installed onto your hard drive.

USER ASSISTANCE AND INFORMATION

John Wiley & Sons, Inc. is pleased to provide assistance to users of this package. Should you have any questions regarding the use of this package, please call our technical support number (212) 850-6194 weekdays between 9 am and 4pm Eastern Standard Time.

To place additional orders or to request information about other Wiley products, please call (800) 879-4539.

CHAPTER 1

BASE PLATE

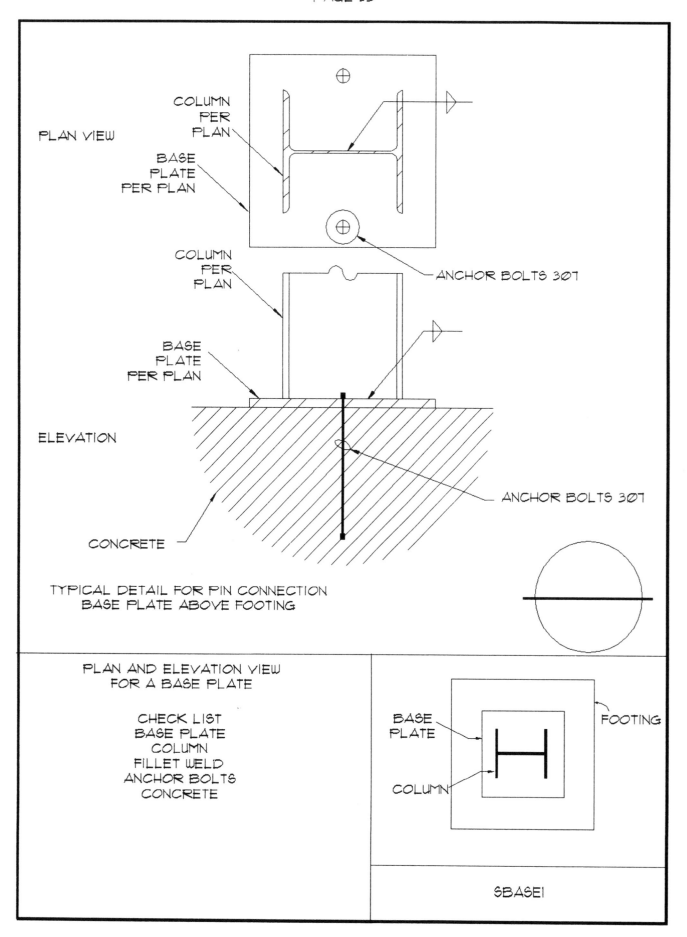

PLAN VIEW

COLUMN
PER
PLAN

BASE
PLATE
PER PLAN

ANCHOR BOLTS 30T

COLUMN
PER
PLAN

BASE
PLATE
PER PLAN

ELEVATION

ANCHOR BOLTS 30T

CONCRETE

TYPICAL DETAIL FOR PIN CONNECTION
BASE PLATE ABOVE FOOTING

PLAN AND ELEVATION VIEW
FOR A BASE PLATE

CHECK LIST
BASE PLATE
COLUMN
FILLET WELD
ANCHOR BOLTS
CONCRETE

BASE
PLATE

FOOTING

COLUMN

SBASE1

PAGE24

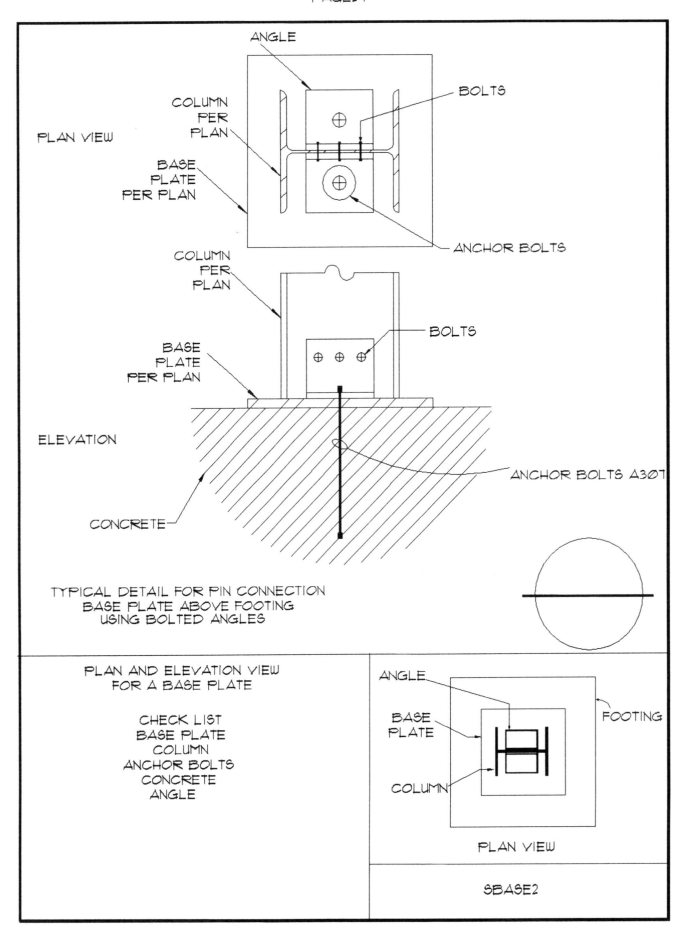

PLAN VIEW

ANGLE

BOLTS

COLUMN
PER
PLAN

BASE
PLATE
PER PLAN

ANCHOR BOLTS

COLUMN
PER
PLAN

BOLTS

BASE
PLATE
PER PLAN

ELEVATION

ANCHOR BOLTS A307

CONCRETE

TYPICAL DETAIL FOR PIN CONNECTION
BASE PLATE ABOVE FOOTING
USING BOLTED ANGLES

PLAN AND ELEVATION VIEW
FOR A BASE PLATE

CHECK LIST
BASE PLATE
COLUMN
ANCHOR BOLTS
CONCRETE
ANGLE

ANGLE

BASE
PLATE

FOOTING

COLUMN

PLAN VIEW

SBASE2

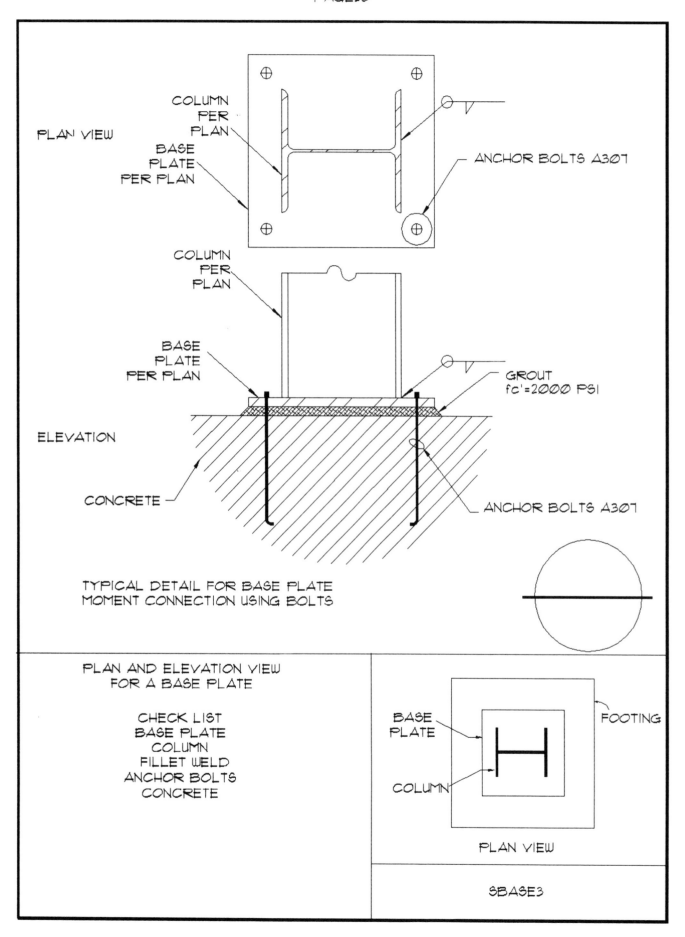

PLAN VIEW

COLUMN
PER
PLAN

BASE
PLATE
PER PLAN

ANCHOR BOLTS A307

COLUMN
PER
PLAN

BASE
PLATE
PER PLAN

GROUT
fc'=2000 PSI

ELEVATION

CONCRETE

ANCHOR BOLTS A307

TYPICAL DETAIL FOR BASE PLATE
MOMENT CONNECTION USING BOLTS

PLAN AND ELEVATION VIEW
FOR A BASE PLATE

CHECK LIST
BASE PLATE
COLUMN
FILLET WELD
ANCHOR BOLTS
CONCRETE

BASE
PLATE

FOOTING

COLUMN

PLAN VIEW

SBASE3

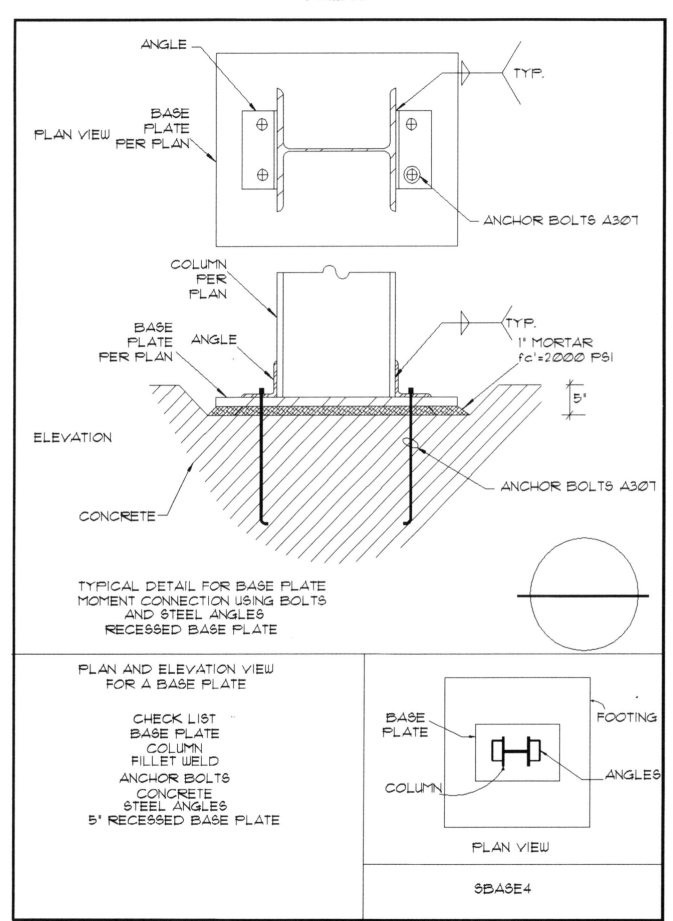

PLAN VIEW

ANGLE

BASE
PLATE
PER PLAN

TYP.

ANCHOR BOLTS A307

COLUMN
PER
PLAN

BASE
PLATE
PER PLAN

ANGLE

TYP.
1" MORTAR
fc'=2000 PSI

5"

ELEVATION

CONCRETE

ANCHOR BOLTS A307

TYPICAL DETAIL FOR BASE PLATE
MOMENT CONNECTION USING BOLTS
AND STEEL ANGLES
RECESSED BASE PLATE

PLAN AND ELEVATION VIEW
FOR A BASE PLATE

CHECK LIST
BASE PLATE
COLUMN
FILLET WELD
ANCHOR BOLTS
CONCRETE
STEEL ANGLES
5" RECESSED BASE PLATE

BASE
PLATE

FOOTING

COLUMN

ANGLES

PLAN VIEW

SBASE4

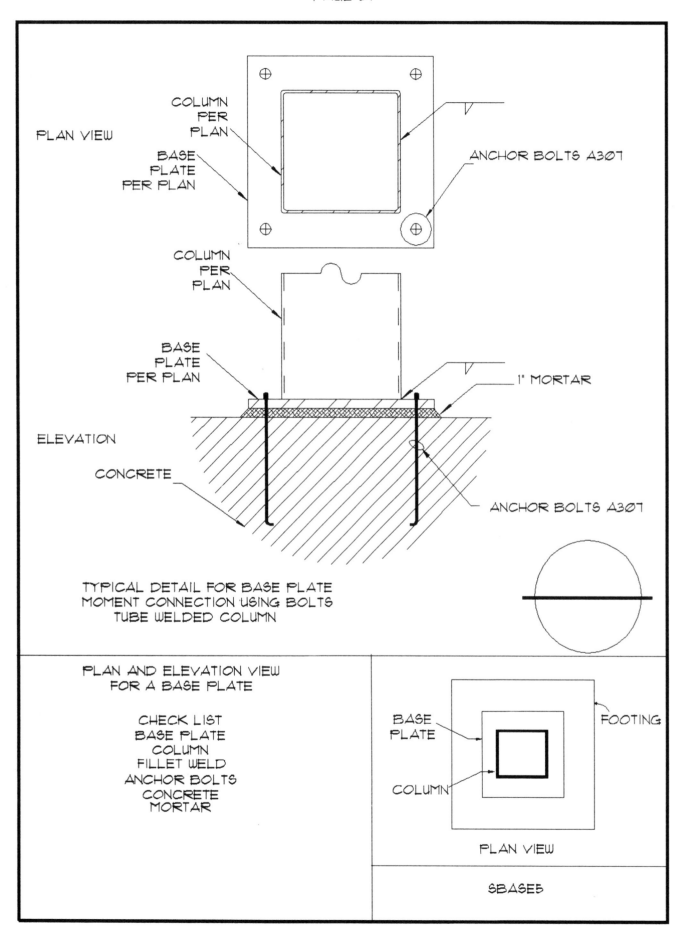

PLAN VIEW

COLUMN
PER
PLAN

BASE
PLATE
PER PLAN

ANCHOR BOLTS A307

COLUMN
PER
PLAN

BASE
PLATE
PER PLAN

1" MORTAR

ELEVATION

CONCRETE

ANCHOR BOLTS A307

TYPICAL DETAIL FOR BASE PLATE
MOMENT CONNECTION USING BOLTS
TUBE WELDED COLUMN

PLAN AND ELEVATION VIEW
FOR A BASE PLATE

CHECK LIST
BASE PLATE
COLUMN
FILLET WELD
ANCHOR BOLTS
CONCRETE
MORTAR

BASE
PLATE

FOOTING

COLUMN

PLAN VIEW

SBASE5

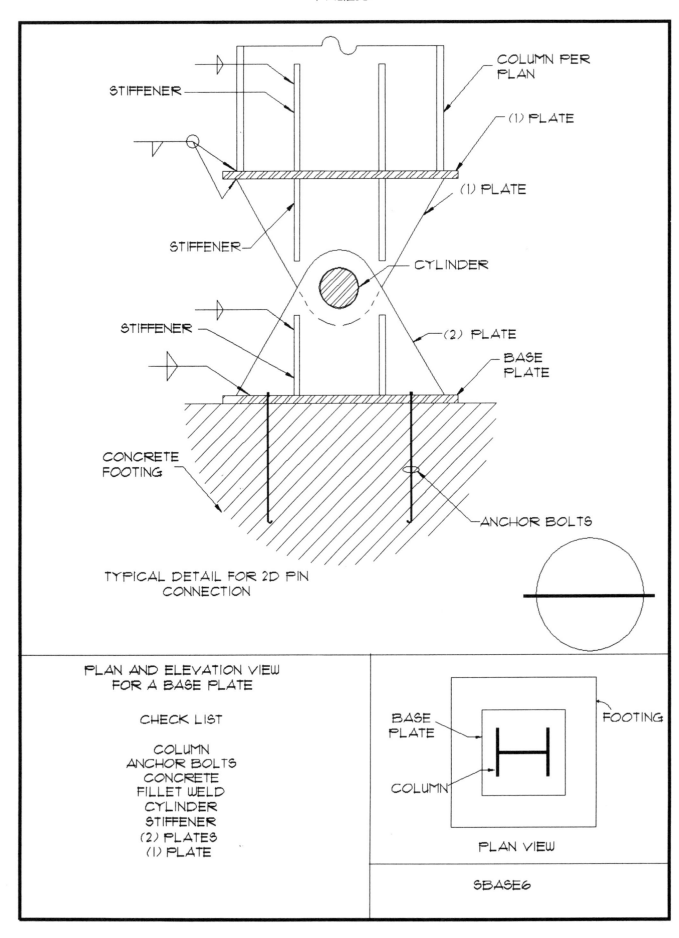

STIFFENER

COLUMN PER PLAN

(1) PLATE

(1) PLATE

STIFFENER

CYLINDER

STIFFENER

(2) PLATE

BASE PLATE

CONCRETE FOOTING

ANCHOR BOLTS

TYPICAL DETAIL FOR 2D PIN CONNECTION

PLAN AND ELEVATION VIEW FOR A BASE PLATE

CHECK LIST

COLUMN
ANCHOR BOLTS
CONCRETE
FILLET WELD
CYLINDER
STIFFENER
(2) PLATES
(1) PLATE

BASE PLATE

FOOTING

COLUMN

PLAN VIEW

SBASE6

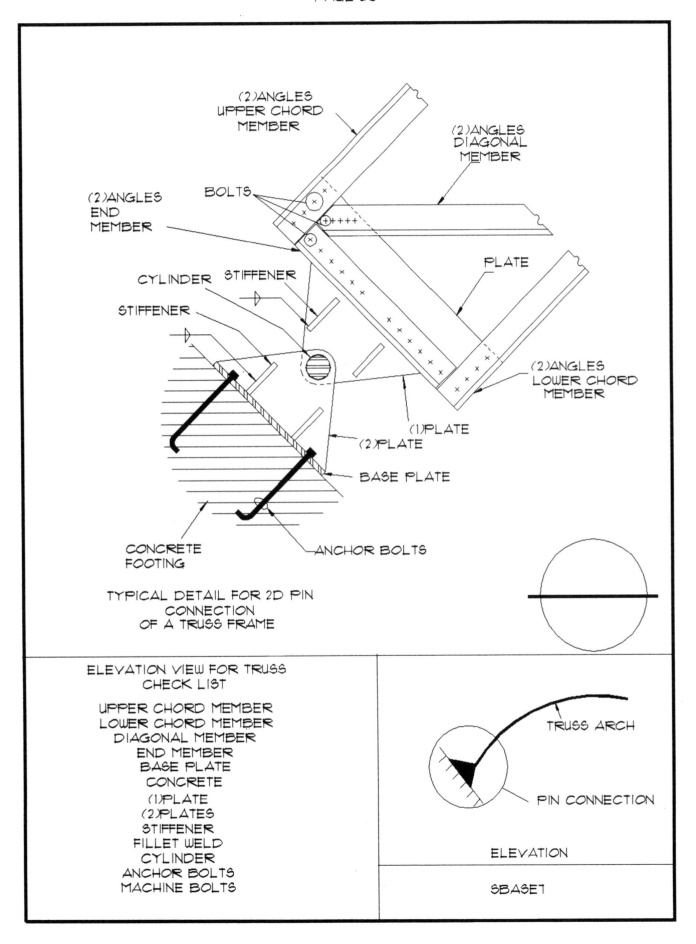

(2)ANGLES
UPPER CHORD
MEMBER

(2)ANGLES
DIAGONAL
MEMBER

BOLTS

(2)ANGLES
END
MEMBER

PLATE

CYLINDER   STIFFENER

STIFFENER

(2)ANGLES
LOWER CHORD
MEMBER

(1)PLATE

(2)PLATE

BASE PLATE

CONCRETE
FOOTING

ANCHOR BOLTS

TYPICAL DETAIL FOR 2D PIN
CONNECTION
OF A TRUSS FRAME

ELEVATION VIEW FOR TRUSS
CHECK LIST

UPPER CHORD MEMBER
LOWER CHORD MEMBER
DIAGONAL MEMBER
END MEMBER
BASE PLATE
CONCRETE
(1)PLATE
(2)PLATES
STIFFENER
FILLET WELD
CYLINDER
ANCHOR BOLTS
MACHINE BOLTS

TRUSS ARCH

PIN CONNECTION

ELEVATION

SBASE1

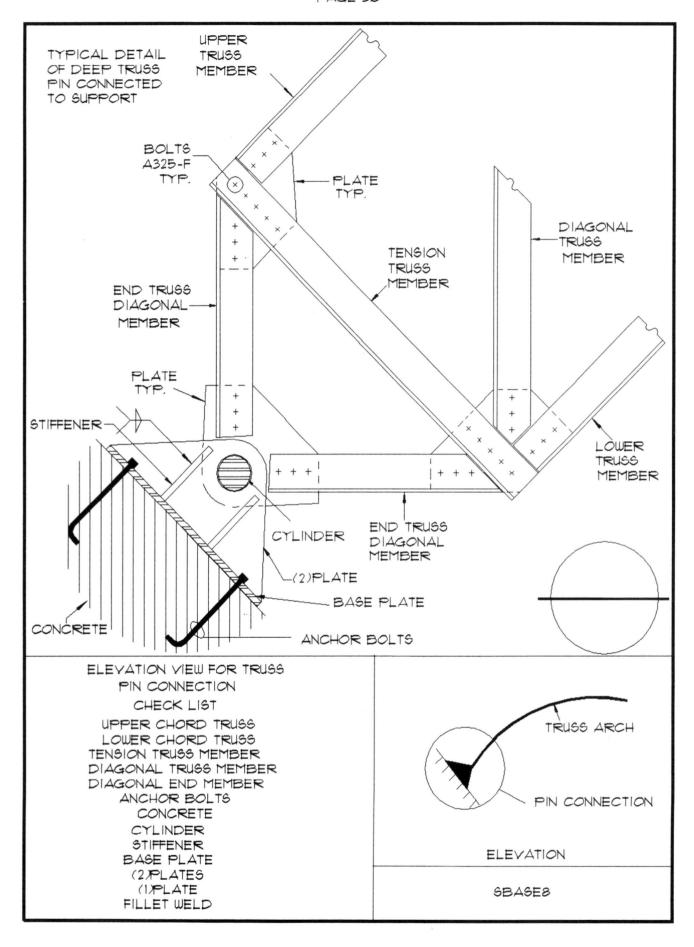

TYPICAL DETAIL OF DEEP TRUSS PIN CONNECTED TO SUPPORT

UPPER TRUSS MEMBER

BOLTS A325-F TYP.

PLATE TYP.

DIAGONAL TRUSS MEMBER

TENSION TRUSS MEMBER

END TRUSS DIAGONAL MEMBER

PLATE TYP.

STIFFENER

LOWER TRUSS MEMBER

CYLINDER

END TRUSS DIAGONAL MEMBER

(2)PLATE

BASE PLATE

CONCRETE

ANCHOR BOLTS

ELEVATION VIEW FOR TRUSS PIN CONNECTION
CHECK LIST
UPPER CHORD TRUSS
LOWER CHORD TRUSS
TENSION TRUSS MEMBER
DIAGONAL TRUSS MEMBER
DIAGONAL END MEMBER
ANCHOR BOLTS
CONCRETE
CYLINDER
STIFFENER
BASE PLATE
(2)PLATES
(1)PLATE
FILLET WELD

TRUSS ARCH

PIN CONNECTION

ELEVATION

SBASE8

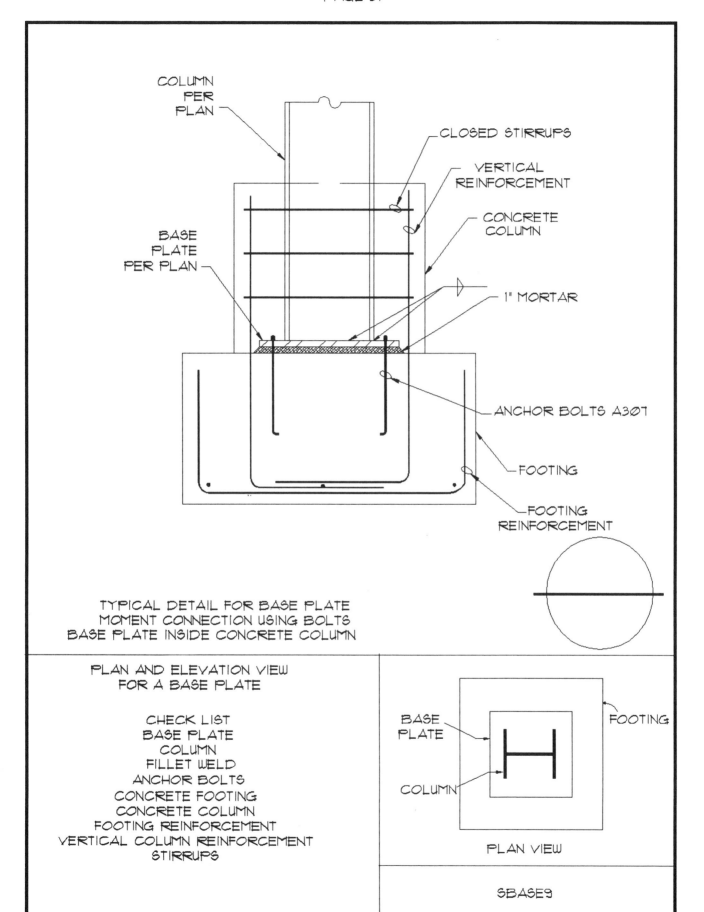

COLUMN
PER
PLAN

CLOSED STIRRUPS

VERTICAL
REINFORCEMENT

CONCRETE
COLUMN

BASE
PLATE
PER PLAN

1" MORTAR

ANCHOR BOLTS A307

FOOTING

FOOTING
REINFORCEMENT

TYPICAL DETAIL FOR BASE PLATE
MOMENT CONNECTION USING BOLTS
BASE PLATE INSIDE CONCRETE COLUMN

PLAN AND ELEVATION VIEW
FOR A BASE PLATE

CHECK LIST
BASE PLATE
COLUMN
FILLET WELD
ANCHOR BOLTS
CONCRETE FOOTING
CONCRETE COLUMN
FOOTING REINFORCEMENT
VERTICAL COLUMN REINFORCEMENT
STIRRUPS

BASE
PLATE

FOOTING

COLUMN

PLAN VIEW

SBASE9

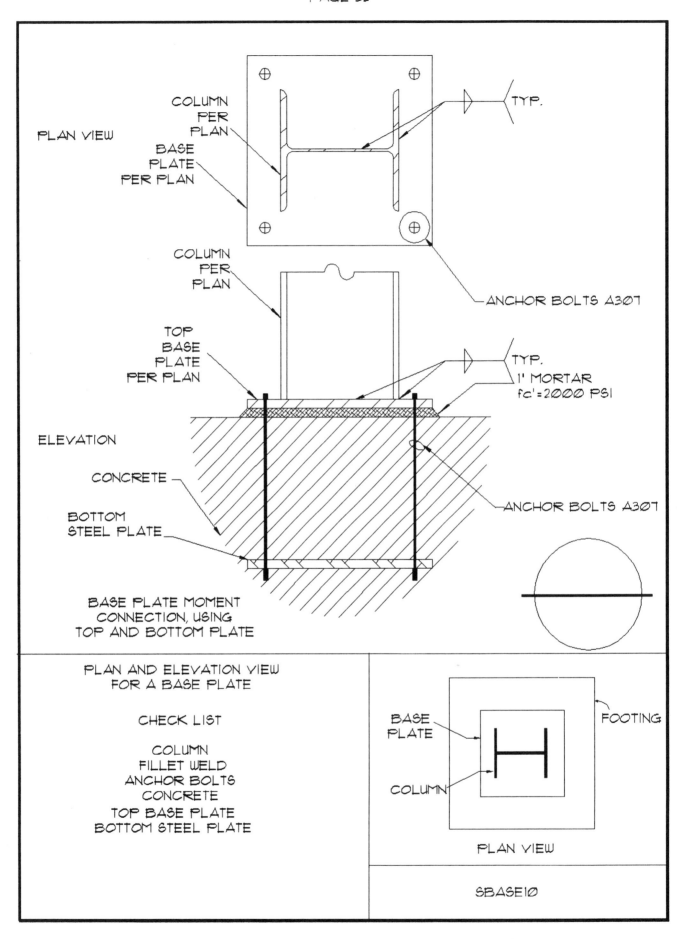

PLAN VIEW

COLUMN
PER
PLAN

BASE
PLATE
PER PLAN

TYP.

ANCHOR BOLTS A307

COLUMN
PER
PLAN

TOP
BASE
PLATE
PER PLAN

TYP.
1' MORTAR
fc'=2000 PSI

ELEVATION

CONCRETE

BOTTOM
STEEL PLATE

ANCHOR BOLTS A307

BASE PLATE MOMENT
CONNECTION, USING
TOP AND BOTTOM PLATE

PLAN AND ELEVATION VIEW
FOR A BASE PLATE

CHECK LIST

COLUMN
FILLET WELD
ANCHOR BOLTS
CONCRETE
TOP BASE PLATE
BOTTOM STEEL PLATE

BASE
PLATE

FOOTING

COLUMN

PLAN VIEW

SBASE10

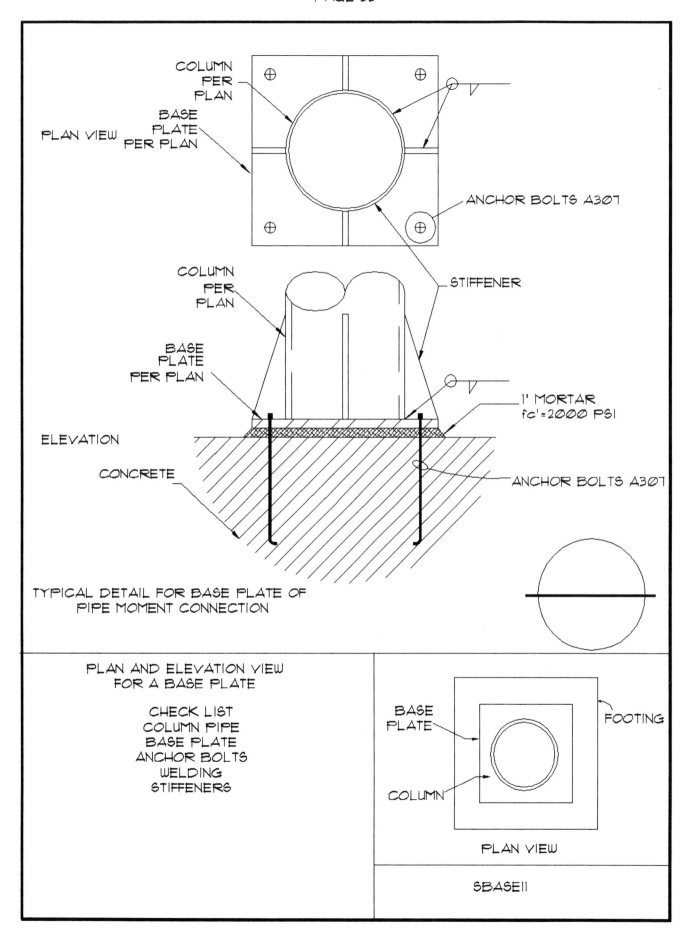

PLAN VIEW

COLUMN
PER
PLAN

BASE
PLATE
PER PLAN

ANCHOR BOLTS A307

STIFFENER

COLUMN
PER
PLAN

BASE
PLATE
PER PLAN

ELEVATION

1' MORTAR
fc'=2000 PSI

CONCRETE

ANCHOR BOLTS A307

TYPICAL DETAIL FOR BASE PLATE OF
PIPE MOMENT CONNECTION

PLAN AND ELEVATION VIEW
FOR A BASE PLATE

CHECK LIST
COLUMN PIPE
BASE PLATE
ANCHOR BOLTS
WELDING
STIFFENERS

BASE
PLATE

FOOTING

COLUMN

PLAN VIEW

SBASE11

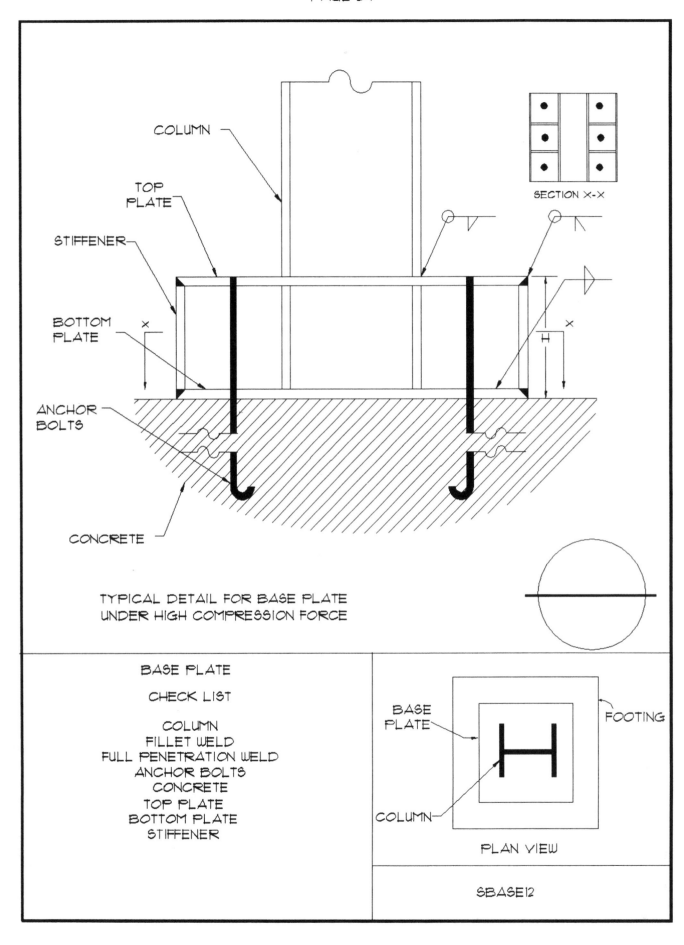

COLUMN

TOP PLATE

STIFFENER

BOTTOM PLATE

ANCHOR BOLTS

CONCRETE

SECTION X-X

TYPICAL DETAIL FOR BASE PLATE
UNDER HIGH COMPRESSION FORCE

BASE PLATE

CHECK LIST

COLUMN
FILLET WELD
FULL PENETRATION WELD
ANCHOR BOLTS
CONCRETE
TOP PLATE
BOTTOM PLATE
STIFFENER

BASE PLATE

FOOTING

COLUMN

PLAN VIEW

SBASE12

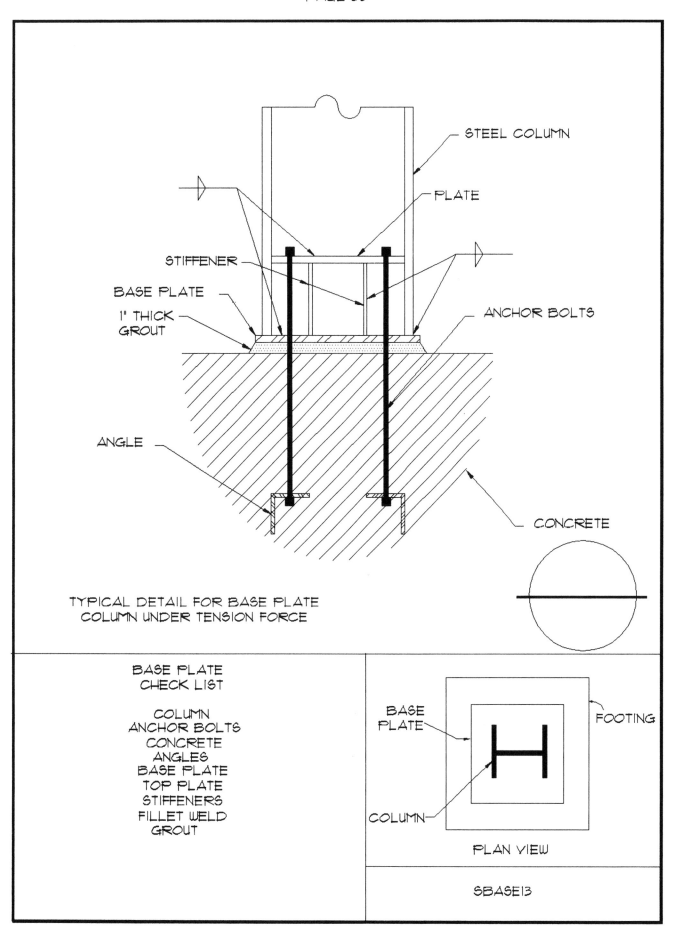

STEEL COLUMN

PLATE

STIFFENER

BASE PLATE

1" THICK GROUT

ANCHOR BOLTS

ANGLE

CONCRETE

TYPICAL DETAIL FOR BASE PLATE
COLUMN UNDER TENSION FORCE

BASE PLATE
CHECK LIST

COLUMN
ANCHOR BOLTS
CONCRETE
ANGLES
BASE PLATE
TOP PLATE
STIFFENERS
FILLET WELD
GROUT

BASE PLATE

FOOTING

COLUMN

PLAN VIEW

SBASE13

CHAPTER 2

LATERAL BRACE
STEEL FRAME

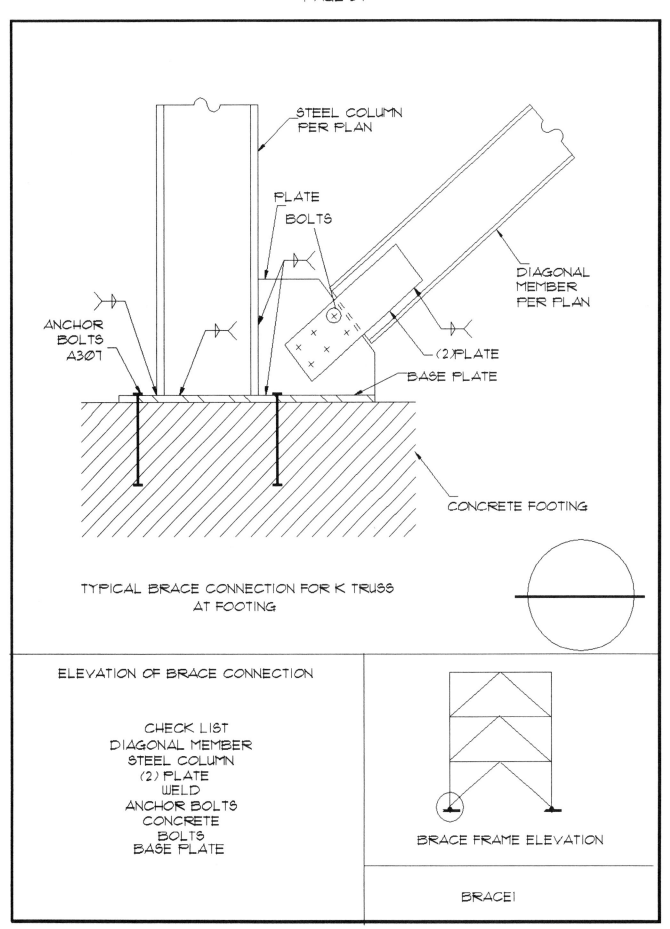

STEEL COLUMN
PER PLAN

PLATE

BOLTS

DIAGONAL
MEMBER
PER PLAN

ANCHOR
BOLTS
A307

(2) PLATE

BASE PLATE

CONCRETE FOOTING

TYPICAL BRACE CONNECTION FOR K TRUSS
AT FOOTING

ELEVATION OF BRACE CONNECTION

CHECK LIST
DIAGONAL MEMBER
STEEL COLUMN
(2) PLATE
WELD
ANCHOR BOLTS
CONCRETE
BOLTS
BASE PLATE

BRACE FRAME ELEVATION

BRACE1

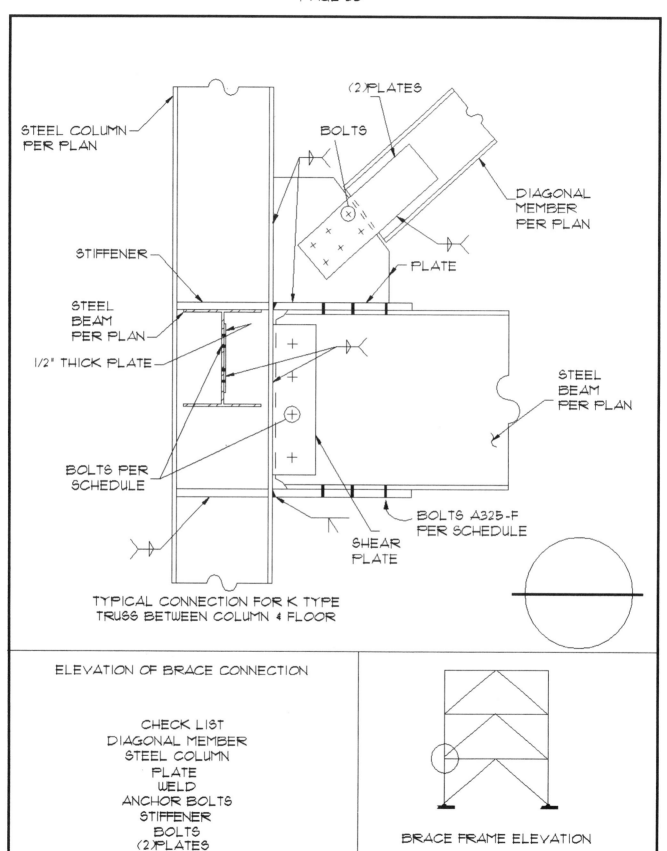

STEEL COLUMN
PER PLAN

(2) PLATES

BOLTS

DIAGONAL
MEMBER
PER PLAN

STIFFENER

PLATE

STEEL
BEAM
PER PLAN

1/2" THICK PLATE

STEEL
BEAM
PER PLAN

BOLTS PER
SCHEDULE

BOLTS A325-F
PER SCHEDULE

SHEAR
PLATE

TYPICAL CONNECTION FOR K TYPE
TRUSS BETWEEN COLUMN & FLOOR

ELEVATION OF BRACE CONNECTION

CHECK LIST
DIAGONAL MEMBER
STEEL COLUMN
PLATE
WELD
ANCHOR BOLTS
STIFFENER
BOLTS
(2) PLATES
SHEAR PLATE

BRACE FRAME ELEVATION

BRACE2

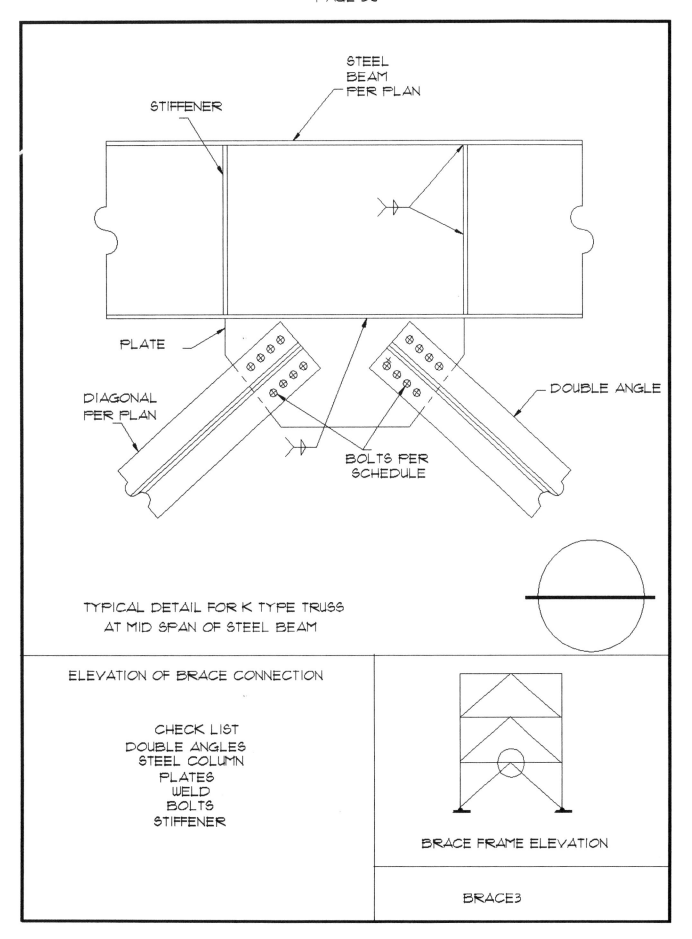

STEEL
BEAM
PER PLAN

STIFFENER

PLATE

DIAGONAL
PER PLAN

DOUBLE ANGLE

BOLTS PER
SCHEDULE

TYPICAL DETAIL FOR K TYPE TRUSS
AT MID SPAN OF STEEL BEAM

ELEVATION OF BRACE CONNECTION

CHECK LIST
DOUBLE ANGLES
STEEL COLUMN
PLATES
WELD
BOLTS
STIFFENER

BRACE FRAME ELEVATION

BRACE3

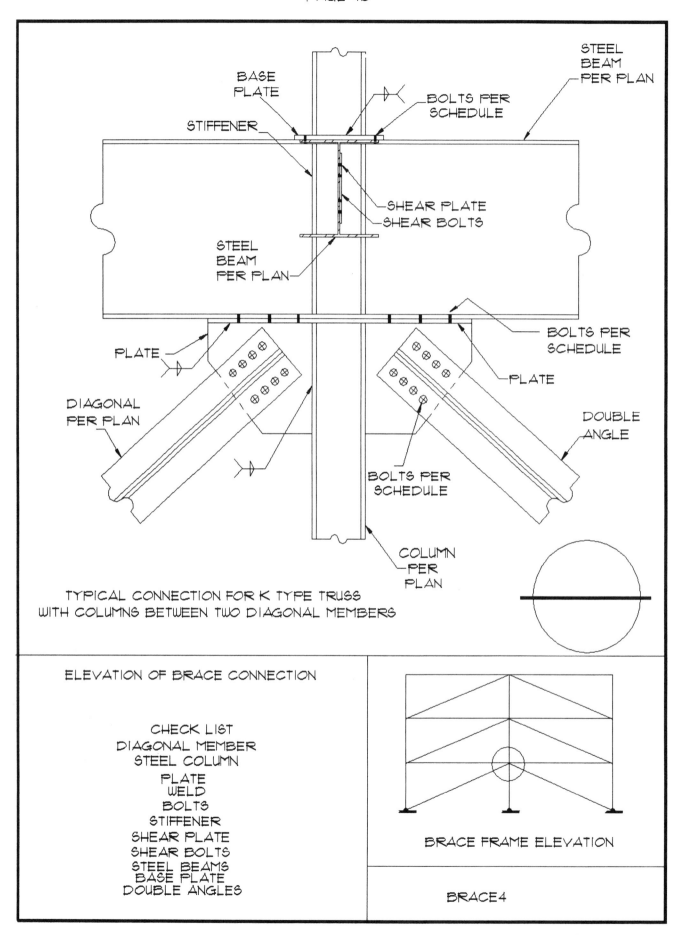

BASE PLATE

STIFFENER

BOLTS PER SCHEDULE

STEEL BEAM PER PLAN

SHEAR PLATE
SHEAR BOLTS

STEEL BEAM PER PLAN

BOLTS PER SCHEDULE

PLATE

PLATE

DIAGONAL PER PLAN

DOUBLE ANGLE

BOLTS PER SCHEDULE

COLUMN PER PLAN

TYPICAL CONNECTION FOR K TYPE TRUSS
WITH COLUMNS BETWEEN TWO DIAGONAL MEMBERS

ELEVATION OF BRACE CONNECTION

CHECK LIST
DIAGONAL MEMBER
STEEL COLUMN
PLATE
WELD
BOLTS
STIFFENER
SHEAR PLATE
SHEAR BOLTS
STEEL BEAMS
BASE PLATE
DOUBLE ANGLES

BRACE FRAME ELEVATION

BRACE4

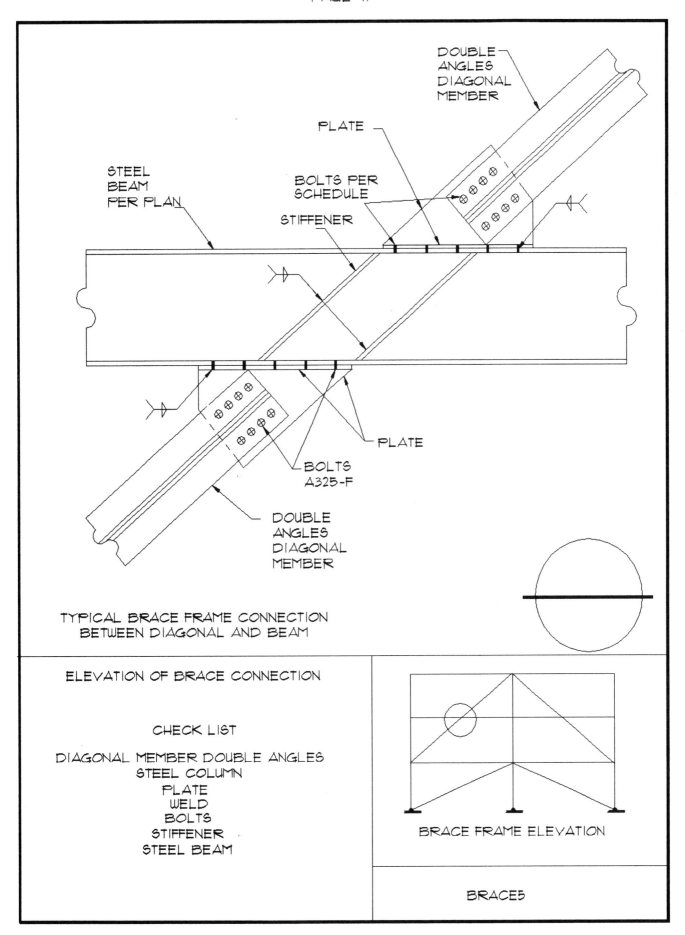

DOUBLE
ANGLES
DIAGONAL
MEMBER

PLATE

STEEL
BEAM
PER PLAN

BOLTS PER
SCHEDULE

STIFFENER

PLATE

BOLTS
A325-F

DOUBLE
ANGLES
DIAGONAL
MEMBER

TYPICAL BRACE FRAME CONNECTION
BETWEEN DIAGONAL AND BEAM

ELEVATION OF BRACE CONNECTION

CHECK LIST

DIAGONAL MEMBER DOUBLE ANGLES
STEEL COLUMN
PLATE
WELD
BOLTS
STIFFENER
STEEL BEAM

BRACE FRAME ELEVATION

BRACE5

PAGE 42

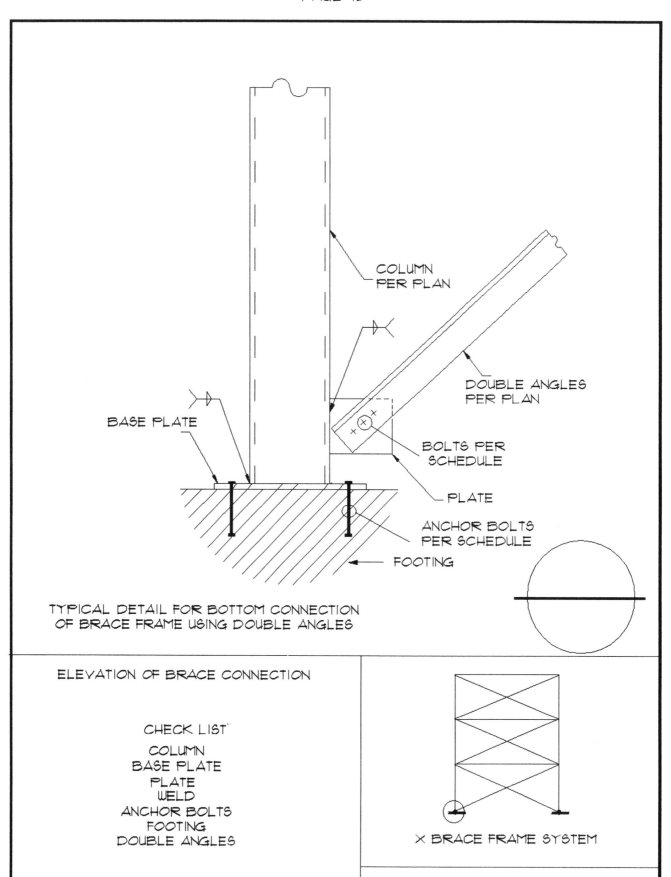

COLUMN
PER PLAN

DOUBLE ANGLES
PER PLAN

BASE PLATE

BOLTS PER
SCHEDULE

PLATE

ANCHOR BOLTS
PER SCHEDULE

FOOTING

TYPICAL DETAIL FOR BOTTOM CONNECTION
OF BRACE FRAME USING DOUBLE ANGLES

ELEVATION OF BRACE CONNECTION

CHECK LIST

COLUMN
BASE PLATE
PLATE
WELD
ANCHOR BOLTS
FOOTING
DOUBLE ANGLES

X BRACE FRAME SYSTEM

BRACE6

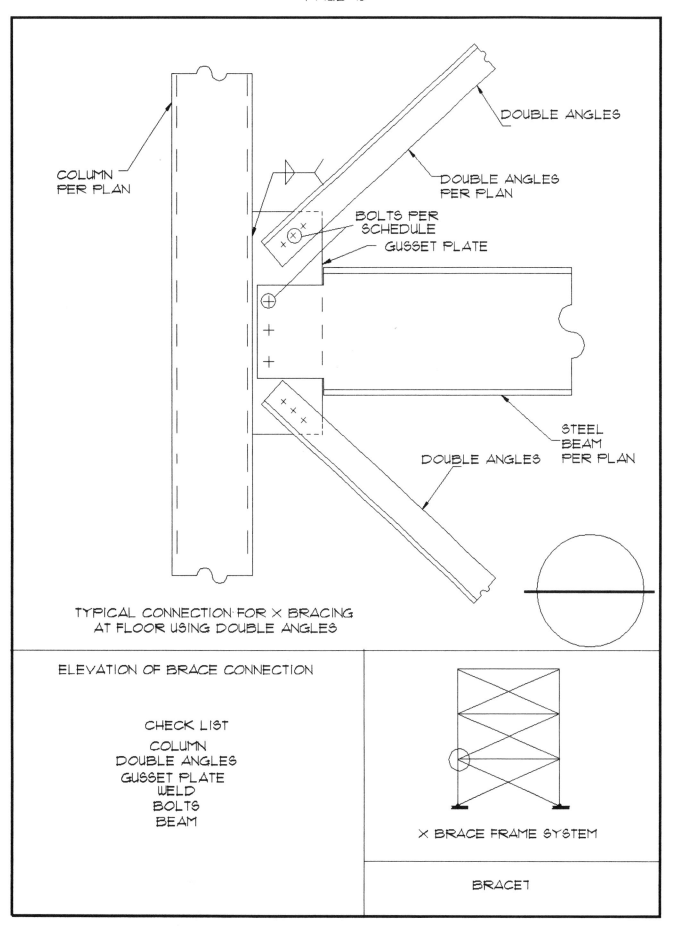

COLUMN
PER PLAN

DOUBLE ANGLES

DOUBLE ANGLES
PER PLAN

BOLTS PER
SCHEDULE

GUSSET PLATE

STEEL
BEAM
PER PLAN

DOUBLE ANGLES

TYPICAL CONNECTION FOR X BRACING
AT FLOOR USING DOUBLE ANGLES

ELEVATION OF BRACE CONNECTION

CHECK LIST

COLUMN
DOUBLE ANGLES
GUSSET PLATE
WELD
BOLTS
BEAM

X BRACE FRAME SYSTEM

BRACE1

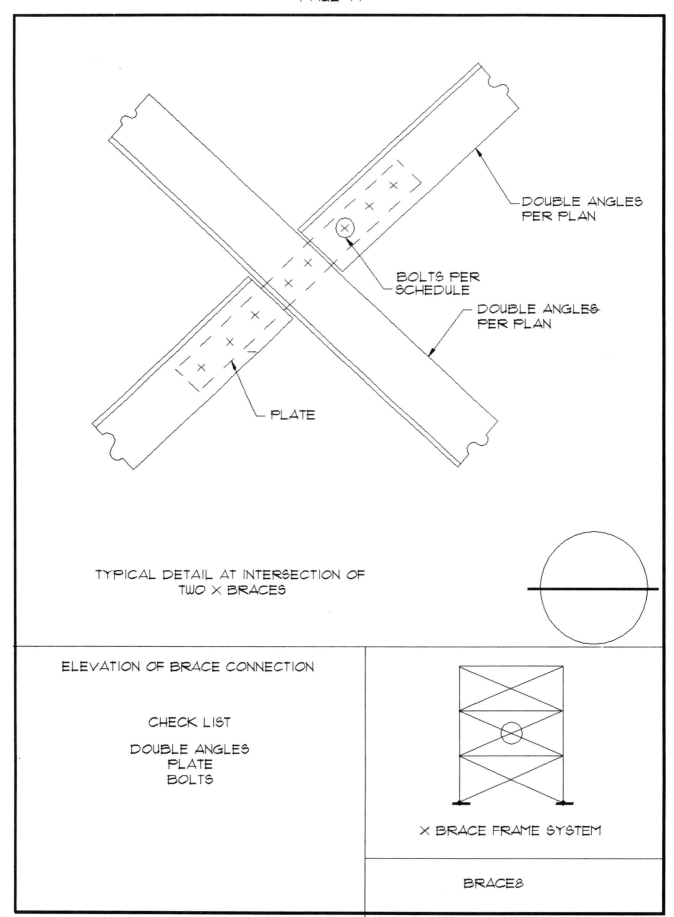

DOUBLE ANGLES
PER PLAN

BOLTS PER
SCHEDULE

DOUBLE ANGLES
PER PLAN

PLATE

TYPICAL DETAIL AT INTERSECTION OF
TWO X BRACES

ELEVATION OF BRACE CONNECTION

CHECK LIST

DOUBLE ANGLES
PLATE
BOLTS

X BRACE FRAME SYSTEM

BRACES

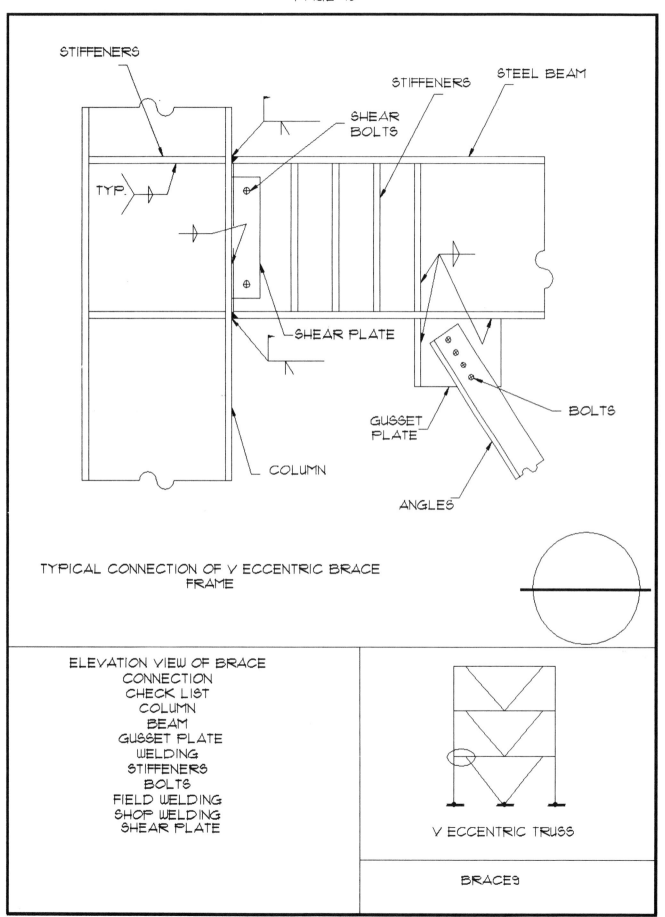

STIFFENERS

TYP.

STIFFENERS

SHEAR BOLTS

STEEL BEAM

SHEAR PLATE

COLUMN

GUSSET PLATE

BOLTS

ANGLES

TYPICAL CONNECTION OF V ECCENTRIC BRACE FRAME

ELEVATION VIEW OF BRACE
CONNECTION
CHECK LIST
COLUMN
BEAM
GUSSET PLATE
WELDING
STIFFENERS
BOLTS
FIELD WELDING
SHOP WELDING
SHEAR PLATE

V ECCENTRIC TRUSS

BRACES

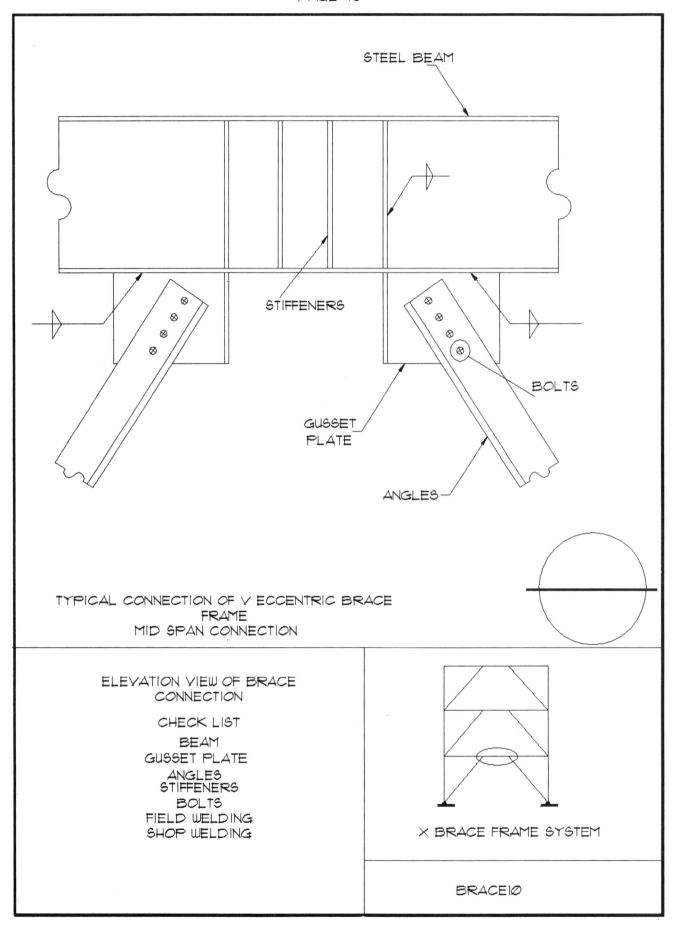

STEEL BEAM

STIFFENERS

BOLTS

GUSSET
PLATE

ANGLES

TYPICAL CONNECTION OF V ECCENTRIC BRACE
FRAME
MID SPAN CONNECTION

ELEVATION VIEW OF BRACE
CONNECTION

CHECK LIST

BEAM
GUSSET PLATE
ANGLES
STIFFENERS
BOLTS
FIELD WELDING
SHOP WELDING

X BRACE FRAME SYSTEM

BRACE10

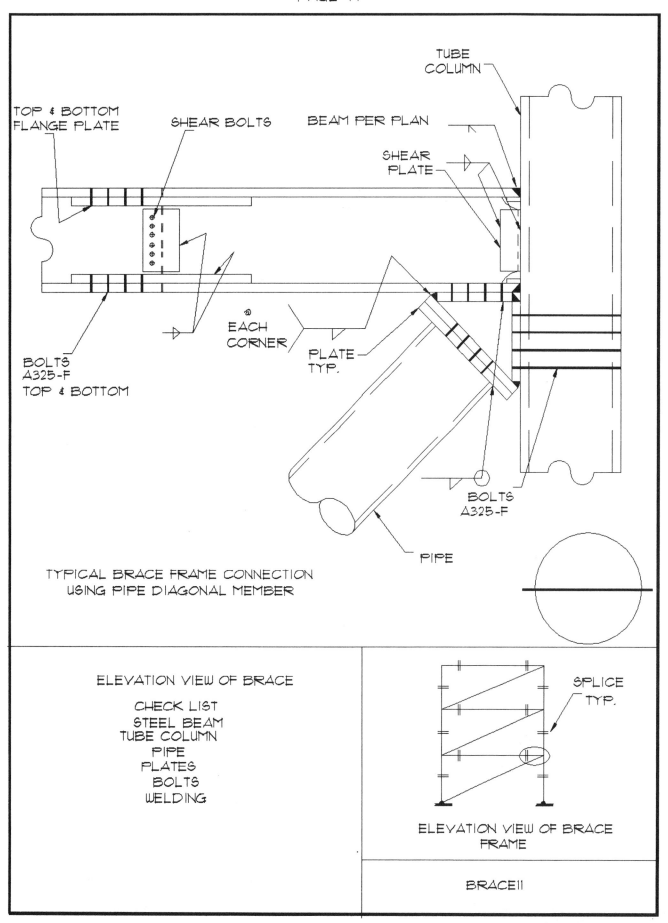

TUBE COLUMN

TOP & BOTTOM FLANGE PLATE

SHEAR BOLTS

BEAM PER PLAN

SHEAR PLATE

BOLTS A325-F TOP & BOTTOM

@ EACH CORNER

PLATE TYP.

BOLTS A325-F

PIPE

TYPICAL BRACE FRAME CONNECTION USING PIPE DIAGONAL MEMBER

ELEVATION VIEW OF BRACE

CHECK LIST
STEEL BEAM
TUBE COLUMN
PIPE
PLATES
BOLTS
WELDING

SPLICE TYP.

ELEVATION VIEW OF BRACE FRAME

BRACE11

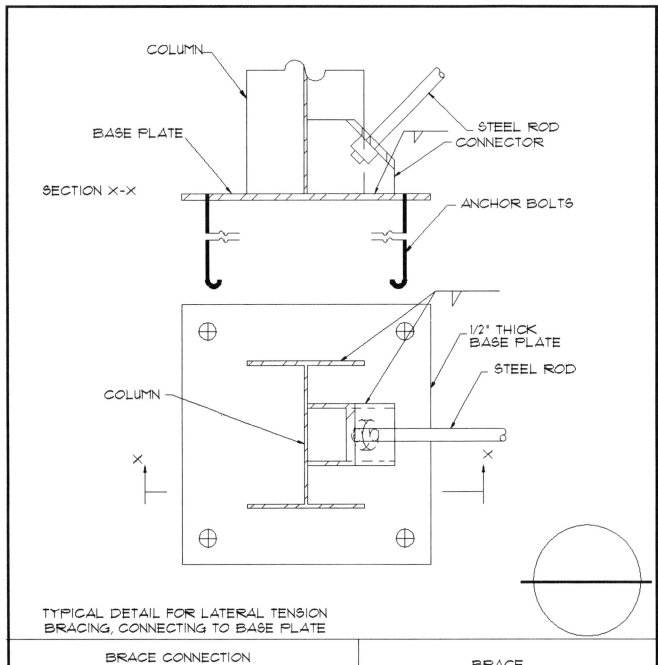

COLUMN

BASE PLATE

SECTION X-X

STEEL ROD

CONNECTOR

ANCHOR BOLTS

1/2" THICK
BASE PLATE

STEEL ROD

COLUMN

TYPICAL DETAIL FOR LATERAL TENSION
BRACING, CONNECTING TO BASE PLATE

BRACE CONNECTION
CHECK LIST
BASE PLATE
COLUMN
WELDING
STEEL CONNECTOR
STEEL ROD
ANCHOR BOLTS

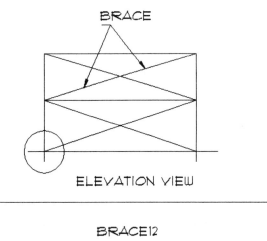

BRACE

ELEVATION VIEW

BRACE12

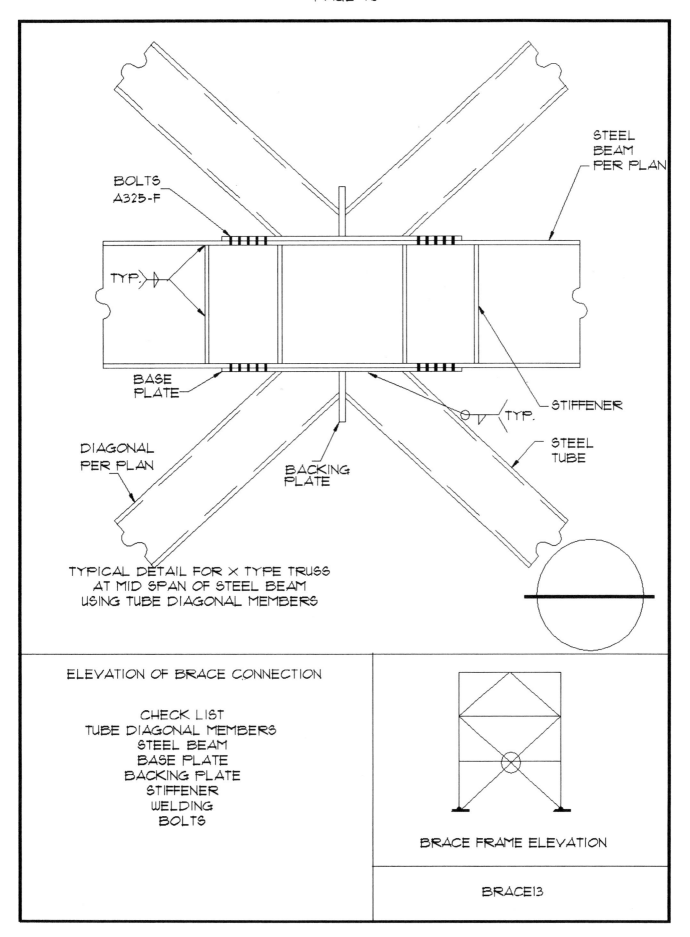

BOLTS
A325-F

STEEL
BEAM
PER PLAN

TYP.

BASE
PLATE

STIFFENER

TYP.

DIAGONAL
PER PLAN

STEEL
TUBE

BACKING
PLATE

TYPICAL DETAIL FOR X TYPE TRUSS
AT MID SPAN OF STEEL BEAM
USING TUBE DIAGONAL MEMBERS

ELEVATION OF BRACE CONNECTION

CHECK LIST
TUBE DIAGONAL MEMBERS
STEEL BEAM
BASE PLATE
BACKING PLATE
STIFFENER
WELDING
BOLTS

BRACE FRAME ELEVATION

BRACE13

CHAPTER 3

BRACKET

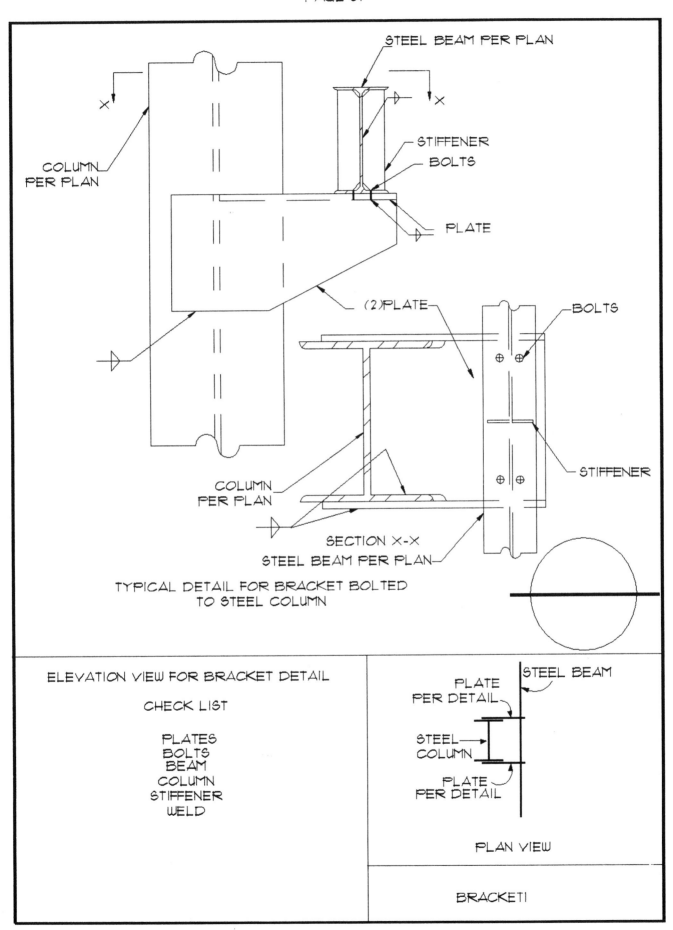

STEEL BEAM PER PLAN

COLUMN
PER PLAN

STIFFENER

BOLTS

PLATE

(2)PLATE

BOLTS

COLUMN
PER PLAN

STIFFENER

SECTION X-X

STEEL BEAM PER PLAN

TYPICAL DETAIL FOR BRACKET BOLTED
TO STEEL COLUMN

ELEVATION VIEW FOR BRACKET DETAIL

CHECK LIST

PLATES
BOLTS
BEAM
COLUMN
STIFFENER
WELD

STEEL BEAM

PLATE
PER DETAIL

STEEL
COLUMN

PLATE
PER DETAIL

PLAN VIEW

BRACKET1

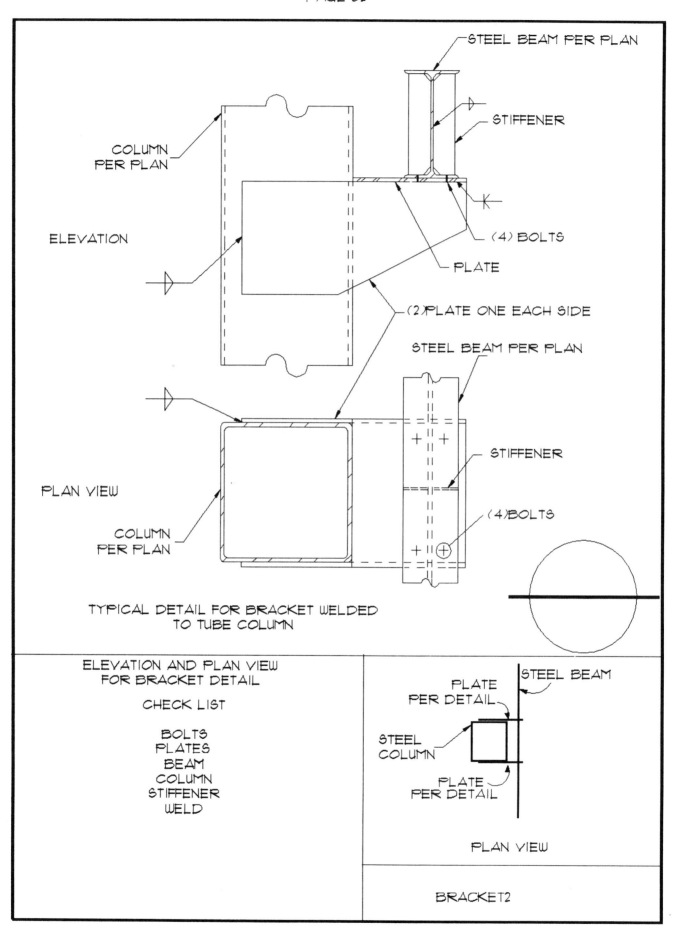

ELEVATION

COLUMN
PER PLAN

STEEL BEAM PER PLAN

STIFFENER

(4) BOLTS

PLATE

(2) PLATE ONE EACH SIDE

STEEL BEAM PER PLAN

STIFFENER

(4) BOLTS

PLAN VIEW

COLUMN
PER PLAN

TYPICAL DETAIL FOR BRACKET WELDED
TO TUBE COLUMN

ELEVATION AND PLAN VIEW
FOR BRACKET DETAIL

CHECK LIST

BOLTS
PLATES
BEAM
COLUMN
STIFFENER
WELD

PLATE
PER DETAIL

STEEL BEAM

STEEL
COLUMN

PLATE
PER DETAIL

PLAN VIEW

BRACKET2

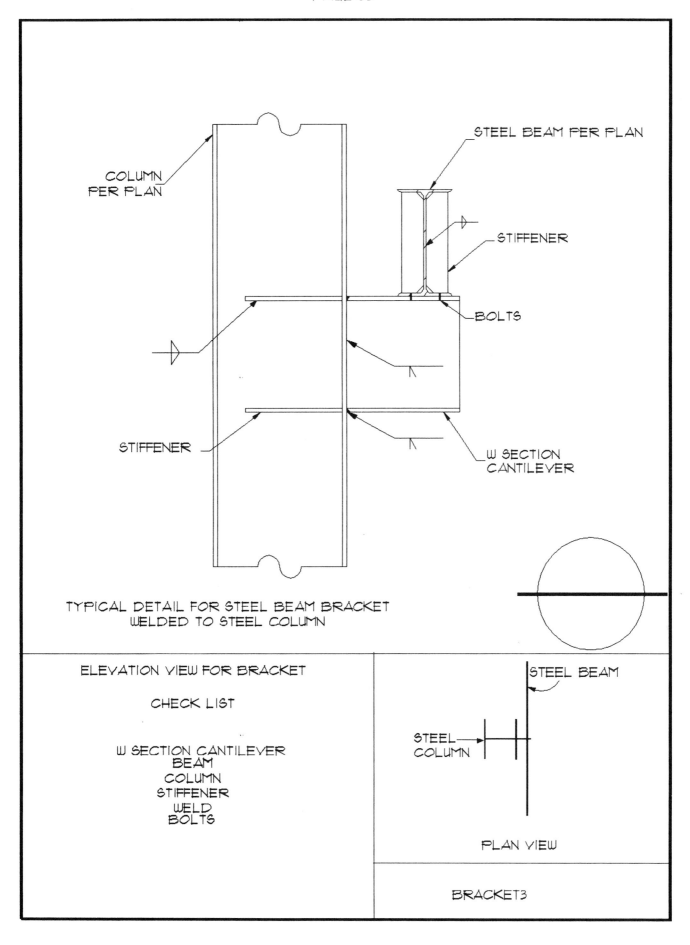

COLUMN
PER PLAN

STEEL BEAM PER PLAN

STIFFENER

BOLTS

STIFFENER

W SECTION
CANTILEVER

TYPICAL DETAIL FOR STEEL BEAM BRACKET
WELDED TO STEEL COLUMN

ELEVATION VIEW FOR BRACKET

CHECK LIST

W SECTION CANTILEVER
BEAM
COLUMN
STIFFENER
WELD
BOLTS

STEEL BEAM

STEEL
COLUMN

PLAN VIEW

BRACKET3

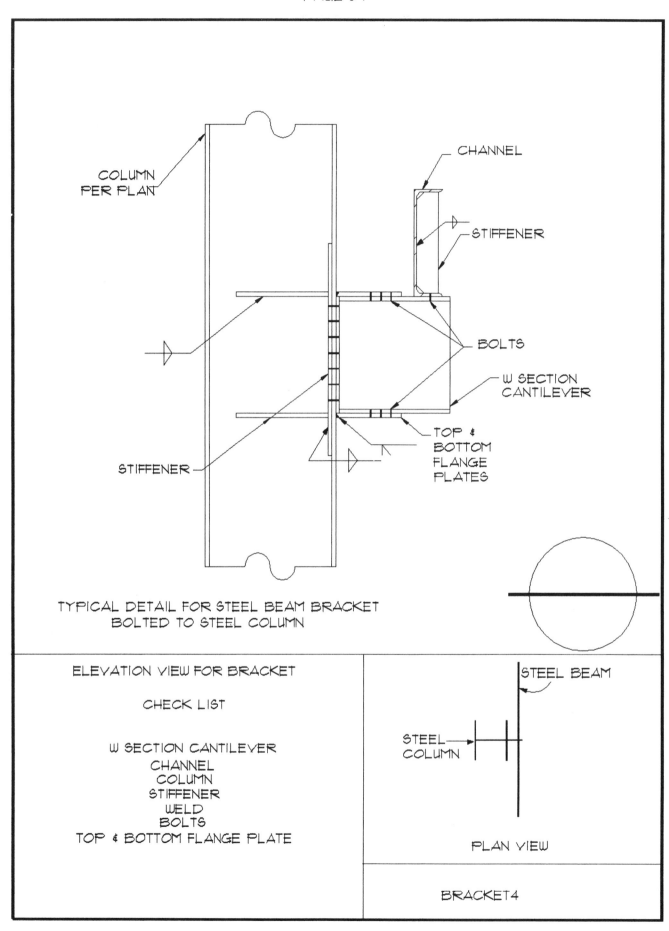

COLUMN
PER PLAN

CHANNEL

STIFFENER

BOLTS

W SECTION
CANTILEVER

TOP &
BOTTOM
FLANGE
PLATES

STIFFENER

TYPICAL DETAIL FOR STEEL BEAM BRACKET
BOLTED TO STEEL COLUMN

ELEVATION VIEW FOR BRACKET

CHECK LIST

W SECTION CANTILEVER
CHANNEL
COLUMN
STIFFENER
WELD
BOLTS
TOP & BOTTOM FLANGE PLATE

STEEL BEAM

STEEL
COLUMN

PLAN VIEW

BRACKET4

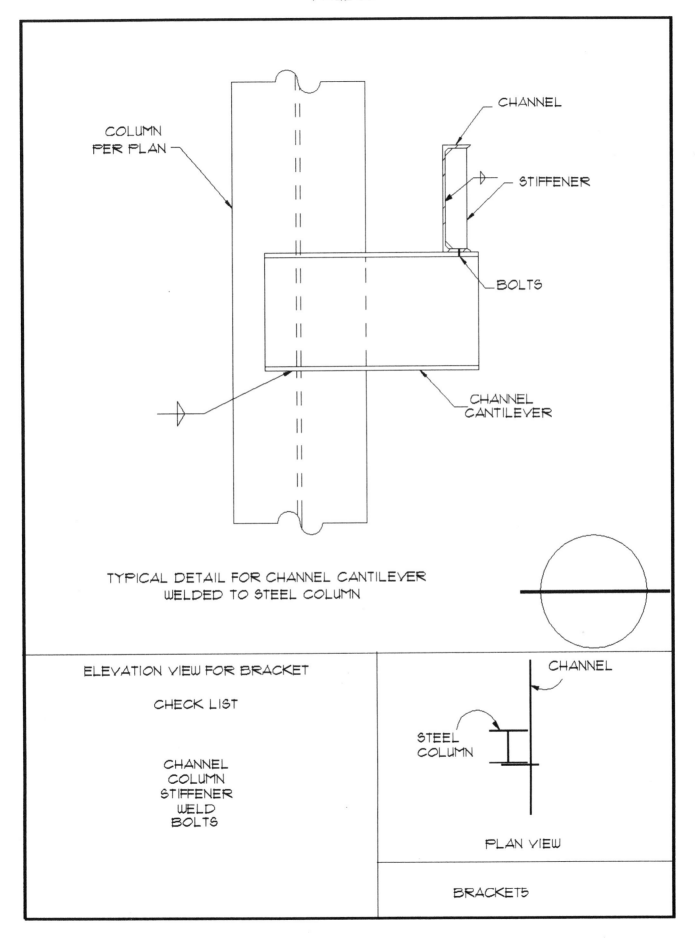

COLUMN
PER PLAN

CHANNEL

STIFFENER

BOLTS

CHANNEL
CANTILEVER

TYPICAL DETAIL FOR CHANNEL CANTILEVER
WELDED TO STEEL COLUMN

ELEVATION VIEW FOR BRACKET

CHECK LIST

CHANNEL
COLUMN
STIFFENER
WELD
BOLTS

CHANNEL

STEEL
COLUMN

PLAN VIEW

BRACKET5

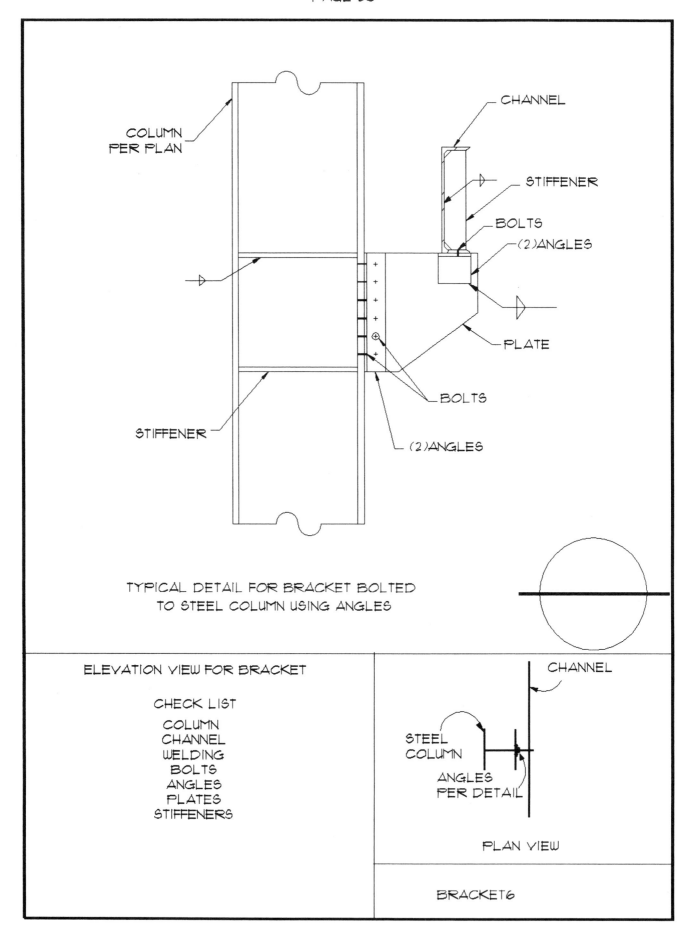

COLUMN
PER PLAN

CHANNEL

STIFFENER

BOLTS

(2)ANGLES

PLATE

BOLTS

(2)ANGLES

STIFFENER

TYPICAL DETAIL FOR BRACKET BOLTED
TO STEEL COLUMN USING ANGLES

ELEVATION VIEW FOR BRACKET

CHECK LIST

COLUMN
CHANNEL
WELDING
BOLTS
ANGLES
PLATES
STIFFENERS

CHANNEL

STEEL
COLUMN

ANGLES
PER DETAIL

PLAN VIEW

BRACKET6

CHAPTER 4

COMPOSITE SECTION STEEL/CONCRETE

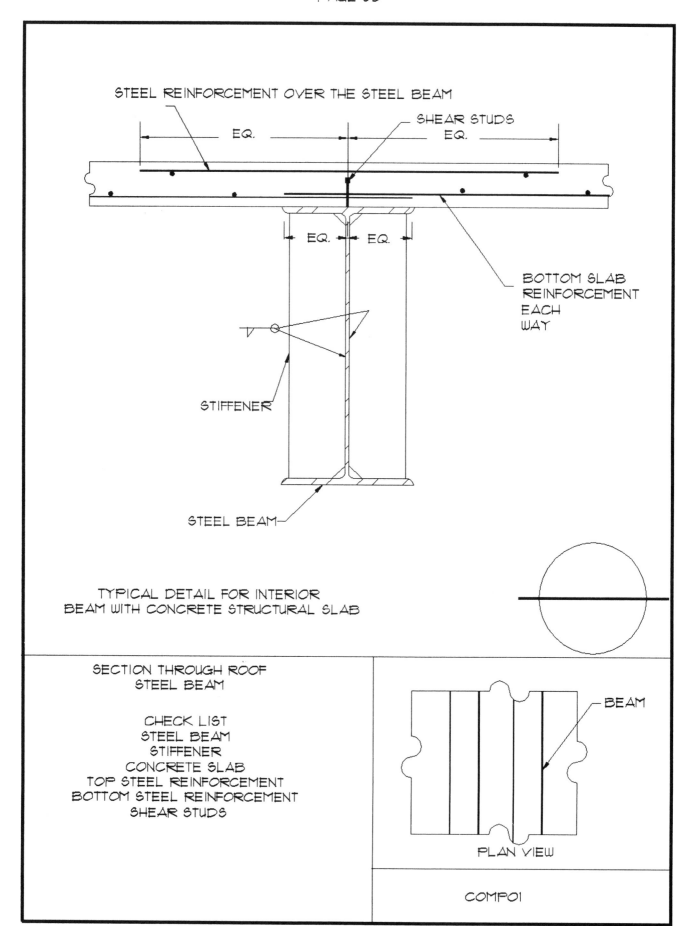

STEEL REINFORCEMENT OVER THE STEEL BEAM

SHEAR STUDS

EQ.

EQ.

EQ.

EQ.

BOTTOM SLAB
REINFORCEMENT
EACH
WAY

STIFFENER

STEEL BEAM

TYPICAL DETAIL FOR INTERIOR
BEAM WITH CONCRETE STRUCTURAL SLAB

SECTION THROUGH ROOF
STEEL BEAM

CHECK LIST
STEEL BEAM
STIFFENER
CONCRETE SLAB
TOP STEEL REINFORCEMENT
BOTTOM STEEL REINFORCEMENT
SHEAR STUDS

BEAM

PLAN VIEW

COMPO1

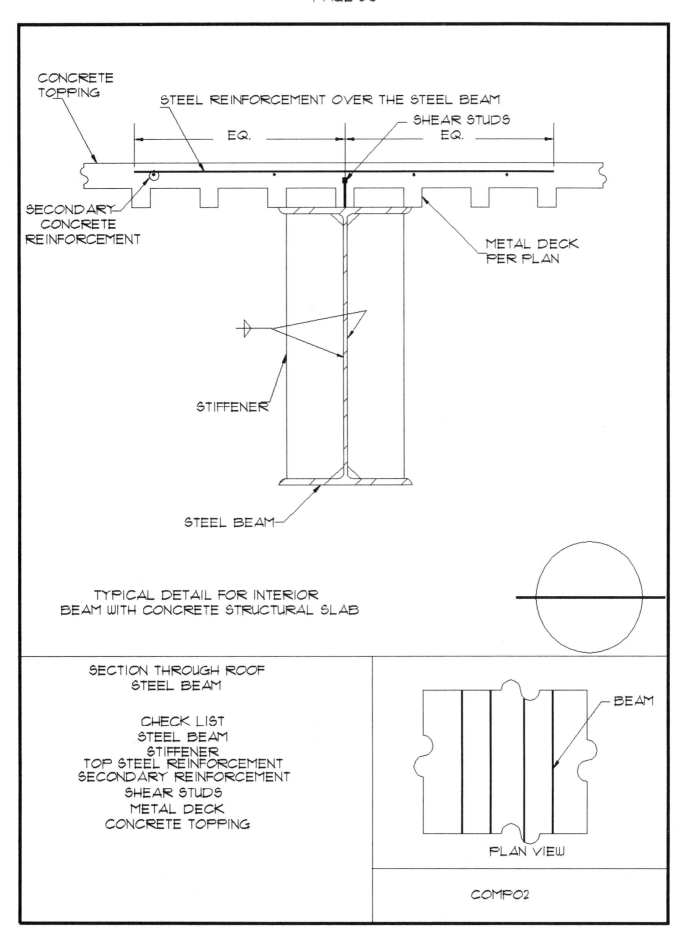

CONCRETE
TOPPING

STEEL REINFORCEMENT OVER THE STEEL BEAM

SHEAR STUDS

EQ.

EQ.

SECONDARY
CONCRETE
REINFORCEMENT

METAL DECK
PER PLAN

STIFFENER

STEEL BEAM

TYPICAL DETAIL FOR INTERIOR
BEAM WITH CONCRETE STRUCTURAL SLAB

SECTION THROUGH ROOF
STEEL BEAM

CHECK LIST
STEEL BEAM
STIFFENER
TOP STEEL REINFORCEMENT
SECONDARY REINFORCEMENT
SHEAR STUDS
METAL DECK
CONCRETE TOPPING

BEAM

PLAN VIEW

COMP02

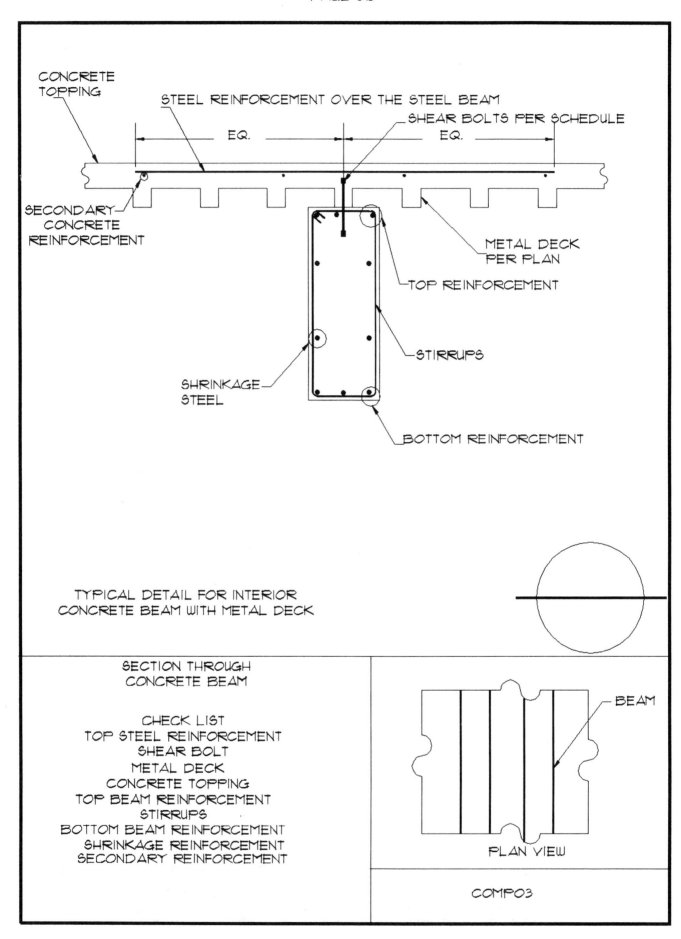

CONCRETE
TOPPING

STEEL REINFORCEMENT OVER THE STEEL BEAM

SHEAR BOLTS PER SCHEDULE

EQ.

EQ.

SECONDARY
CONCRETE
REINFORCEMENT

METAL DECK
PER PLAN

TOP REINFORCEMENT

STIRRUPS

SHRINKAGE
STEEL

BOTTOM REINFORCEMENT

TYPICAL DETAIL FOR INTERIOR
CONCRETE BEAM WITH METAL DECK

SECTION THROUGH
CONCRETE BEAM

CHECK LIST
TOP STEEL REINFORCEMENT
SHEAR BOLT
METAL DECK
CONCRETE TOPPING
TOP BEAM REINFORCEMENT
STIRRUPS
BOTTOM BEAM REINFORCEMENT
SHRINKAGE REINFORCEMENT
SECONDARY REINFORCEMENT

BEAM

PLAN VIEW

COMP03

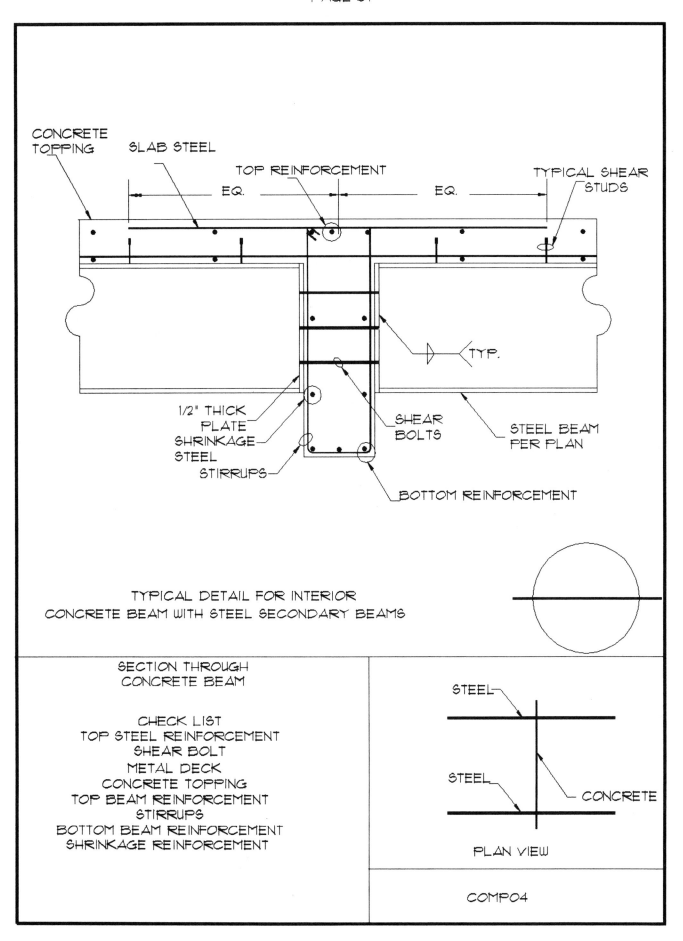

CONCRETE
TOPPING

SLAB STEEL

TOP REINFORCEMENT

TYPICAL SHEAR
STUDS

EQ.

EQ.

1/2" THICK
PLATE
SHRINKAGE
STEEL
STIRRUPS

SHEAR
BOLTS

STEEL BEAM
PER PLAN

TYP.

BOTTOM REINFORCEMENT

TYPICAL DETAIL FOR INTERIOR
CONCRETE BEAM WITH STEEL SECONDARY BEAMS

SECTION THROUGH
CONCRETE BEAM

CHECK LIST
TOP STEEL REINFORCEMENT
SHEAR BOLT
METAL DECK
CONCRETE TOPPING
TOP BEAM REINFORCEMENT
STIRRUPS
BOTTOM BEAM REINFORCEMENT
SHRINKAGE REINFORCEMENT

STEEL

STEEL

CONCRETE

PLAN VIEW

COMP04

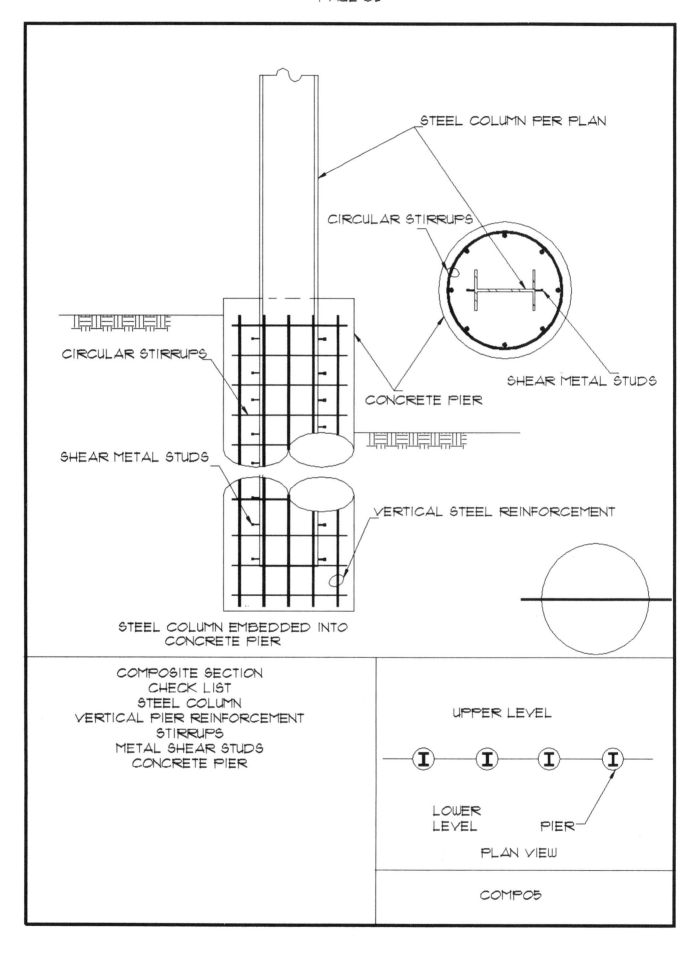

STEEL COLUMN PER PLAN

CIRCULAR STIRRUPS

SHEAR METAL STUDS

CIRCULAR STIRRUPS

CONCRETE PIER

SHEAR METAL STUDS

VERTICAL STEEL REINFORCEMENT

STEEL COLUMN EMBEDDED INTO
CONCRETE PIER

COMPOSITE SECTION
CHECK LIST
STEEL COLUMN
VERTICAL PIER REINFORCEMENT
STIRRUPS
METAL SHEAR STUDS
CONCRETE PIER

UPPER LEVEL

LOWER
LEVEL   PIER

PLAN VIEW

COMPO5

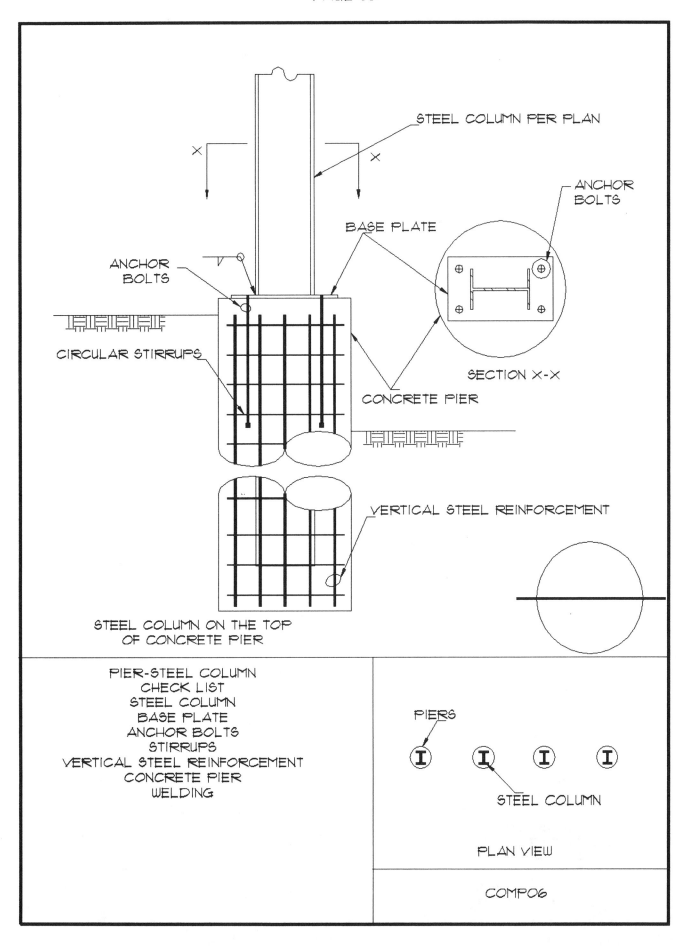

STEEL COLUMN PER PLAN

ANCHOR BOLTS

BASE PLATE

ANCHOR BOLTS

SECTION X-X

CIRCULAR STIRRUPS

CONCRETE PIER

VERTICAL STEEL REINFORCEMENT

STEEL COLUMN ON THE TOP
OF CONCRETE PIER

PIER-STEEL COLUMN
CHECK LIST
STEEL COLUMN
BASE PLATE
ANCHOR BOLTS
STIRRUPS
VERTICAL STEEL REINFORCEMENT
CONCRETE PIER
WELDING

PIERS

STEEL COLUMN

PLAN VIEW

COMP06

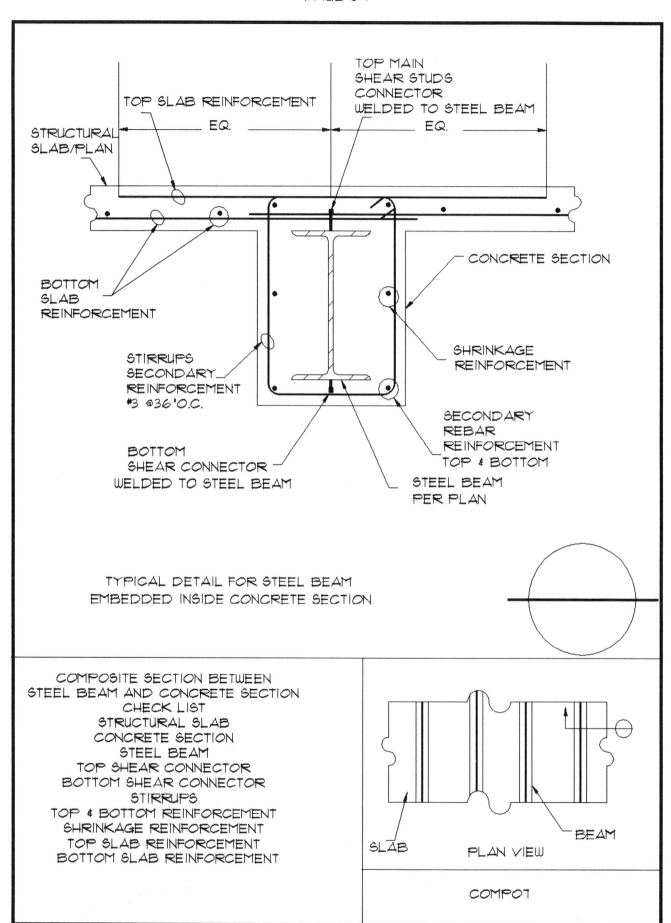

TOP MAIN
SHEAR STUDS
CONNECTOR
WELDED TO STEEL BEAM

TOP SLAB REINFORCEMENT
EQ.

EQ.

STRUCTURAL
SLAB/PLAN

CONCRETE SECTION

BOTTOM
SLAB
REINFORCEMENT

SHRINKAGE
REINFORCEMENT

STIRRUPS
SECONDARY
REINFORCEMENT
#3 @36"O.C.

SECONDARY
REBAR
REINFORCEMENT
TOP & BOTTOM

BOTTOM
SHEAR CONNECTOR
WELDED TO STEEL BEAM

STEEL BEAM
PER PLAN

TYPICAL DETAIL FOR STEEL BEAM
EMBEDDED INSIDE CONCRETE SECTION

COMPOSITE SECTION BETWEEN
STEEL BEAM AND CONCRETE SECTION
CHECK LIST
STRUCTURAL SLAB
CONCRETE SECTION
STEEL BEAM
TOP SHEAR CONNECTOR
BOTTOM SHEAR CONNECTOR
STIRRUPS
TOP & BOTTOM REINFORCEMENT
SHRINKAGE REINFORCEMENT
TOP SLAB REINFORCEMENT
BOTTOM SLAB REINFORCEMENT

SLAB

BEAM

PLAN VIEW

COMPO7

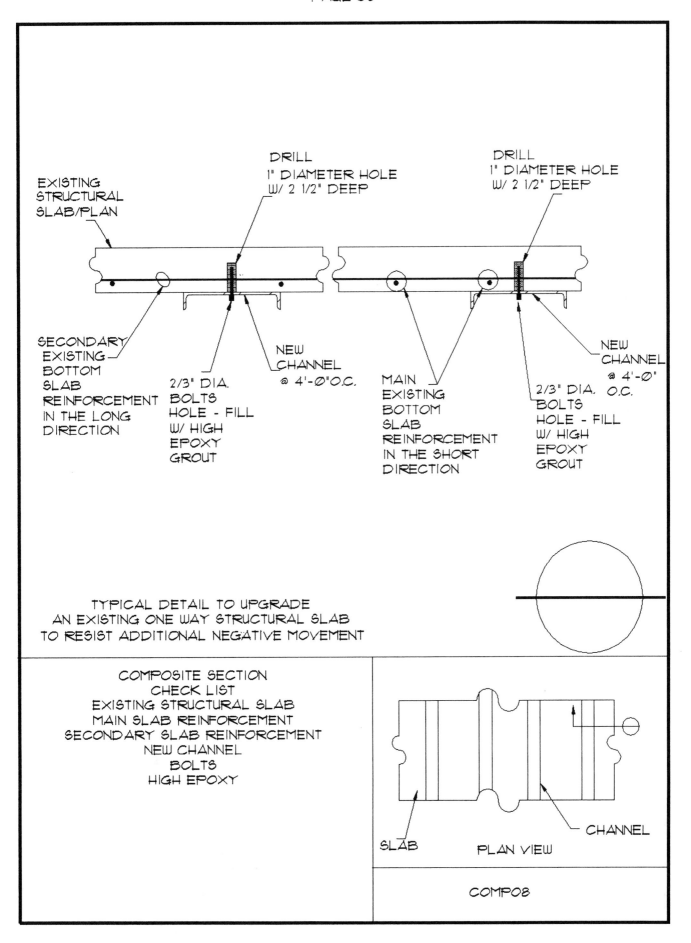

DRILL
1" DIAMETER HOLE
W/ 2 1/2" DEEP

DRILL
1" DIAMETER HOLE
W/ 2 1/2" DEEP

EXISTING
STRUCTURAL
SLAB/PLAN

SECONDARY
EXISTING
BOTTOM
SLAB
REINFORCEMENT
IN THE LONG
DIRECTION

2/3" DIA.
BOLTS
HOLE - FILL
W/ HIGH
EPOXY
GROUT

NEW
CHANNEL
@ 4'-0"O.C.

MAIN
EXISTING
BOTTOM
SLAB
REINFORCEMENT
IN THE SHORT
DIRECTION

2/3" DIA.
BOLTS
HOLE - FILL
W/ HIGH
EPOXY
GROUT

NEW
CHANNEL
@ 4'-0"
O.C.

TYPICAL DETAIL TO UPGRADE
AN EXISTING ONE WAY STRUCTURAL SLAB
TO RESIST ADDITIONAL NEGATIVE MOVEMENT

COMPOSITE SECTION
CHECK LIST
EXISTING STRUCTURAL SLAB
MAIN SLAB REINFORCEMENT
SECONDARY SLAB REINFORCEMENT
NEW CHANNEL
BOLTS
HIGH EPOXY

SLAB

PLAN VIEW

CHANNEL

COMP08

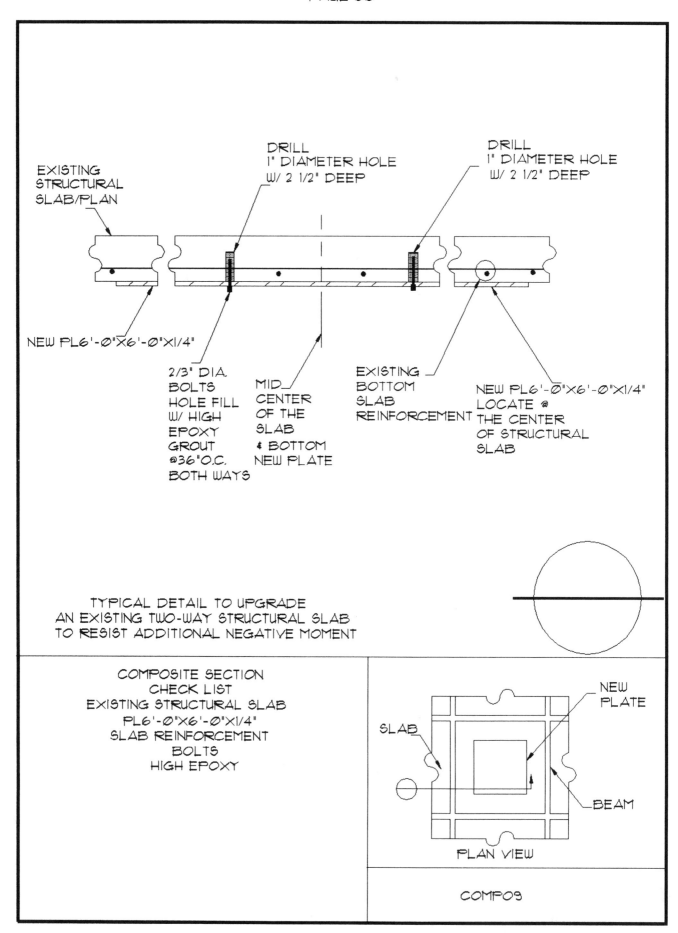

EXISTING
STRUCTURAL
SLAB/PLAN

DRILL
1" DIAMETER HOLE
W/ 2 1/2" DEEP

DRILL
1" DIAMETER HOLE
W/ 2 1/2" DEEP

NEW PL6'-0"X6'-0"X1/4"

2/3" DIA.
BOLTS
HOLE FILL
W/ HIGH
EPOXY
GROUT
@36"O.C.
BOTH WAYS

MID
CENTER
OF THE
SLAB
& BOTTOM
NEW PLATE

EXISTING
BOTTOM
SLAB
REINFORCEMENT

NEW PL6'-0"X6'-0"X1/4"
LOCATE @
THE CENTER
OF STRUCTURAL
SLAB

TYPICAL DETAIL TO UPGRADE
AN EXISTING TWO-WAY STRUCTURAL SLAB
TO RESIST ADDITIONAL NEGATIVE MOMENT

COMPOSITE SECTION
CHECK LIST
EXISTING STRUCTURAL SLAB
PL6'-0"X6'-0"X1/4"
SLAB REINFORCEMENT
BOLTS
HIGH EPOXY

NEW
PLATE

SLAB

BEAM

PLAN VIEW

COMP09

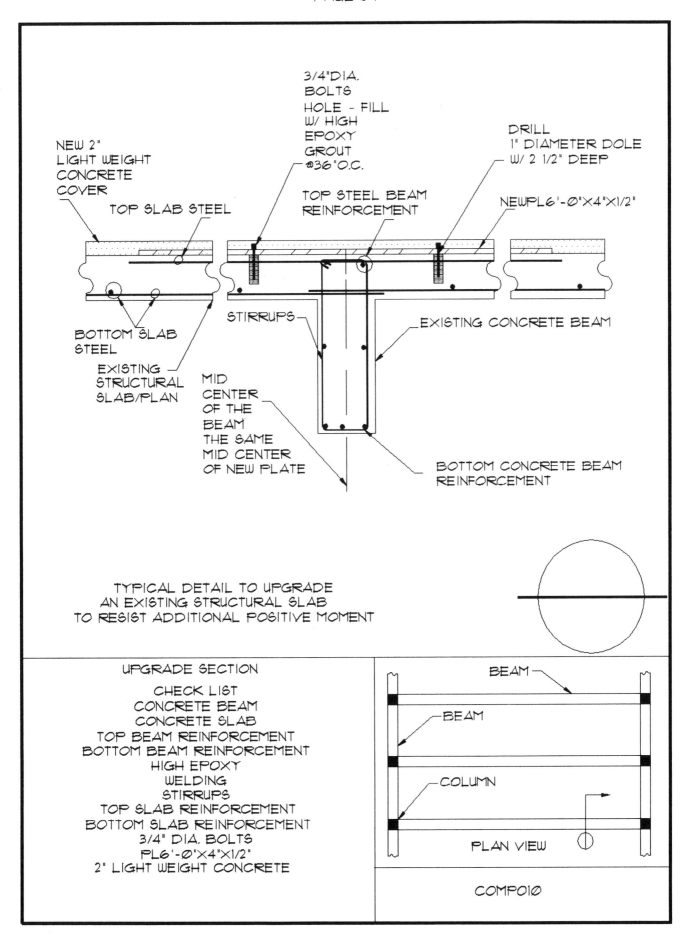

NEW 2"
LIGHT WEIGHT
CONCRETE
COVER

TOP SLAB STEEL

3/4"DIA.
BOLTS
HOLE - FILL
W/ HIGH
EPOXY
GROUT
@36"O.C.

TOP STEEL BEAM
REINFORCEMENT

DRILL
1" DIAMETER DOLE
W/ 2 1/2" DEEP

NEWPL6'-0"X4"X1/2"

BOTTOM SLAB
STEEL

EXISTING
STRUCTURAL
SLAB/PLAN

STIRRUPS

MID
CENTER
OF THE
BEAM
THE SAME
MID CENTER
OF NEW PLATE

EXISTING CONCRETE BEAM

BOTTOM CONCRETE BEAM
REINFORCEMENT

TYPICAL DETAIL TO UPGRADE
AN EXISTING STRUCTURAL SLAB
TO RESIST ADDITIONAL POSITIVE MOMENT

UPGRADE SECTION
CHECK LIST
CONCRETE BEAM
CONCRETE SLAB
TOP BEAM REINFORCEMENT
BOTTOM BEAM REINFORCEMENT
HIGH EPOXY
WELDING
STIRRUPS
TOP SLAB REINFORCEMENT
BOTTOM SLAB REINFORCEMENT
3/4" DIA. BOLTS
PL6'-0'X4"X1/2"
2" LIGHT WEIGHT CONCRETE

BEAM

BEAM

COLUMN

PLAN VIEW

COMP010

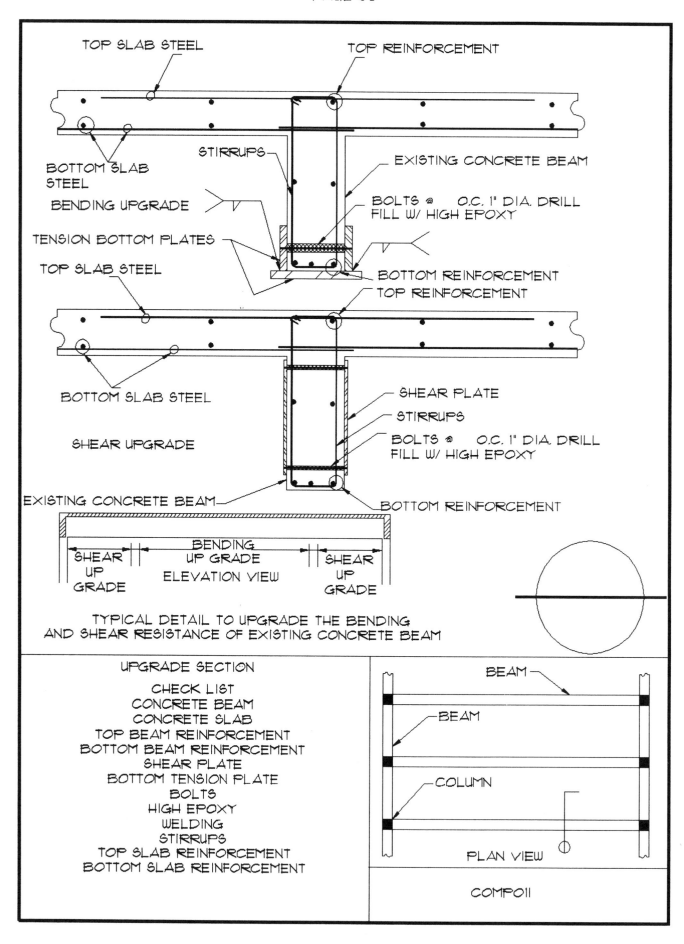

TOP SLAB STEEL

TOP REINFORCEMENT

BOTTOM SLAB STEEL

STIRRUPS

EXISTING CONCRETE BEAM

BENDING UPGRADE

BOLTS @   O.C. 1" DIA. DRILL FILL W/ HIGH EPOXY

TENSION BOTTOM PLATES

BOTTOM REINFORCEMENT

TOP SLAB STEEL

TOP REINFORCEMENT

BOTTOM SLAB STEEL

SHEAR PLATE

STIRRUPS

SHEAR UPGRADE

BOLTS @   O.C. 1" DIA. DRILL FILL W/ HIGH EPOXY

EXISTING CONCRETE BEAM

BOTTOM REINFORCEMENT

SHEAR UP GRADE

BENDING UP GRADE

ELEVATION VIEW

SHEAR UP GRADE

TYPICAL DETAIL TO UPGRADE THE BENDING AND SHEAR RESISTANCE OF EXISTING CONCRETE BEAM

UPGRADE SECTION

CHECK LIST
CONCRETE BEAM
CONCRETE SLAB
TOP BEAM REINFORCEMENT
BOTTOM BEAM REINFORCEMENT
SHEAR PLATE
BOTTOM TENSION PLATE
BOLTS
HIGH EPOXY
WELDING
STIRRUPS
TOP SLAB REINFORCEMENT
BOTTOM SLAB REINFORCEMENT

BEAM

BEAM

COLUMN

PLAN VIEW

COMPO11

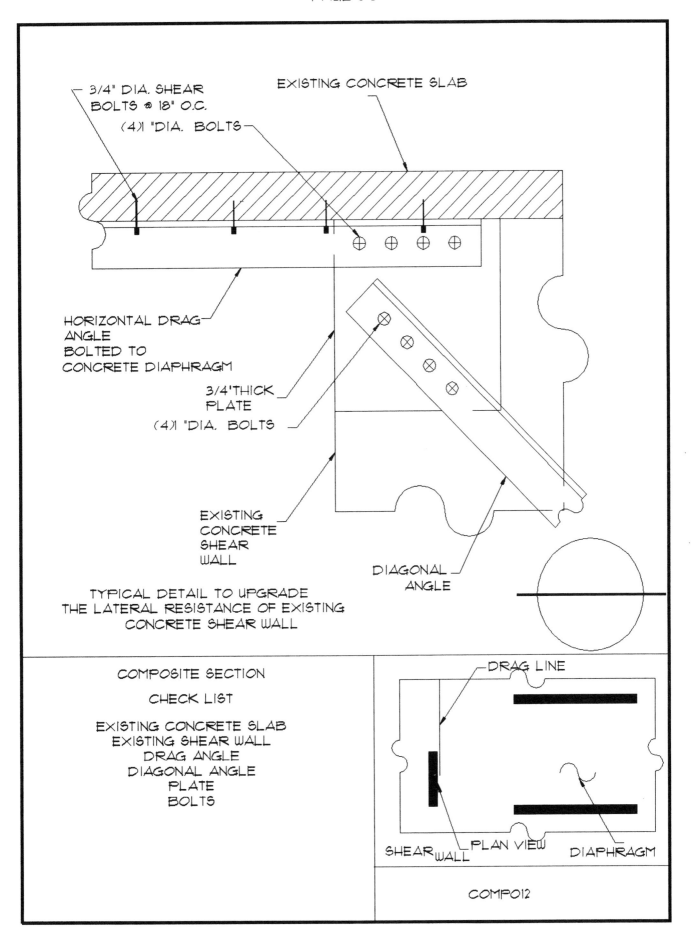

3/4" DIA. SHEAR
BOLTS @ 18" O.C.
(4)1 "DIA. BOLTS

EXISTING CONCRETE SLAB

HORIZONTAL DRAG
ANGLE
BOLTED TO
CONCRETE DIAPHRAGM

3/4'THICK
PLATE
(4)1 "DIA. BOLTS

EXISTING
CONCRETE
SHEAR
WALL

DIAGONAL
ANGLE

TYPICAL DETAIL TO UPGRADE
THE LATERAL RESISTANCE OF EXISTING
CONCRETE SHEAR WALL

COMPOSITE SECTION

CHECK LIST

EXISTING CONCRETE SLAB
EXISTING SHEAR WALL
DRAG ANGLE
DIAGONAL ANGLE
PLATE
BOLTS

DRAG LINE

SHEAR WALL   PLAN VIEW   DIAPHRAGM

COMPO12

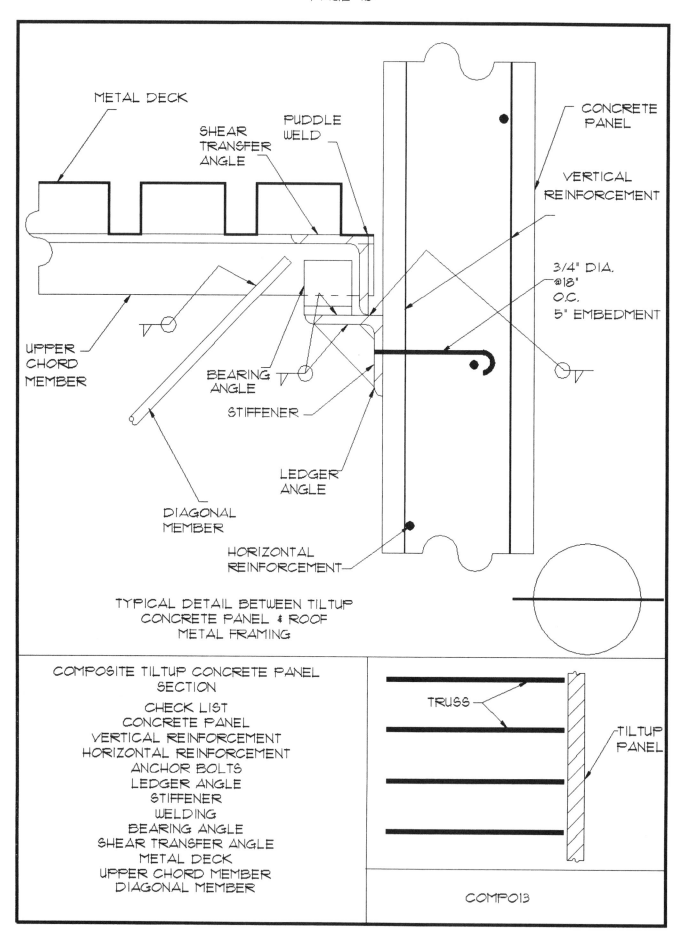

METAL DECK

SHEAR TRANSFER ANGLE

PUDDLE WELD

CONCRETE PANEL

VERTICAL REINFORCEMENT

3/4" DIA. @18" O.C. 5" EMBEDMENT

UPPER CHORD MEMBER

BEARING ANGLE

STIFFENER

LEDGER ANGLE

DIAGONAL MEMBER

HORIZONTAL REINFORCEMENT

TYPICAL DETAIL BETWEEN TILTUP CONCRETE PANEL & ROOF METAL FRAMING

COMPOSITE TILTUP CONCRETE PANEL SECTION

CHECK LIST
CONCRETE PANEL
VERTICAL REINFORCEMENT
HORIZONTAL REINFORCEMENT
ANCHOR BOLTS
LEDGER ANGLE
STIFFENER
WELDING
BEARING ANGLE
SHEAR TRANSFER ANGLE
METAL DECK
UPPER CHORD MEMBER
DIAGONAL MEMBER

TRUSS

TILTUP PANEL

COMPO13

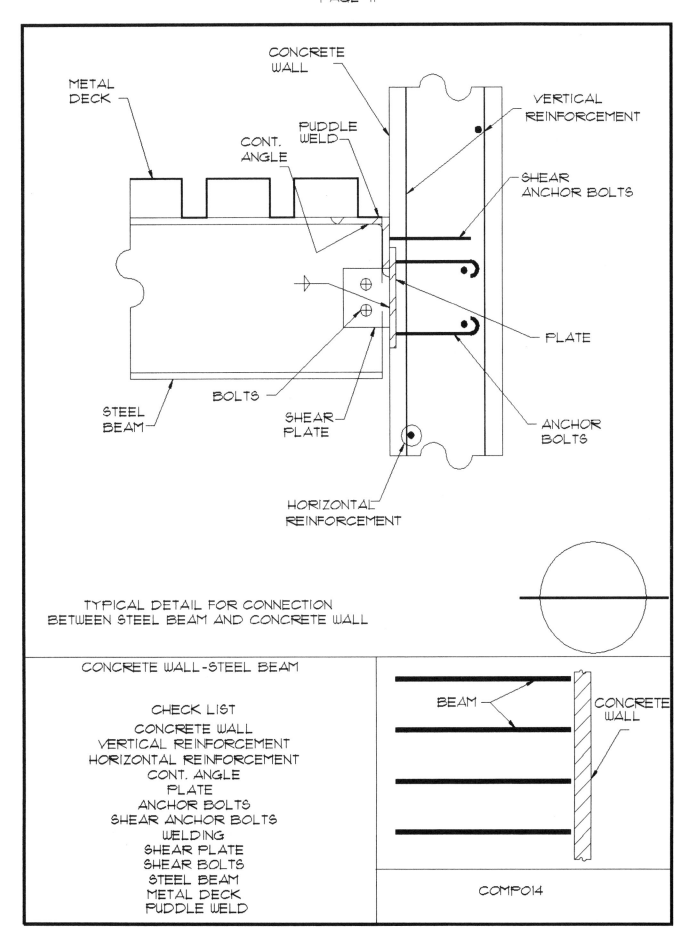

METAL DECK

CONCRETE WALL

CONT. ANGLE

PUDDLE WELD

VERTICAL REINFORCEMENT

SHEAR ANCHOR BOLTS

PLATE

STEEL BEAM

BOLTS

SHEAR PLATE

ANCHOR BOLTS

HORIZONTAL REINFORCEMENT

TYPICAL DETAIL FOR CONNECTION
BETWEEN STEEL BEAM AND CONCRETE WALL

CONCRETE WALL-STEEL BEAM

CHECK LIST
CONCRETE WALL
VERTICAL REINFORCEMENT
HORIZONTAL REINFORCEMENT
CONT. ANGLE
PLATE
ANCHOR BOLTS
SHEAR ANCHOR BOLTS
WELDING
SHEAR PLATE
SHEAR BOLTS
STEEL BEAM
METAL DECK
PUDDLE WELD

BEAM

CONCRETE WALL

COMP014

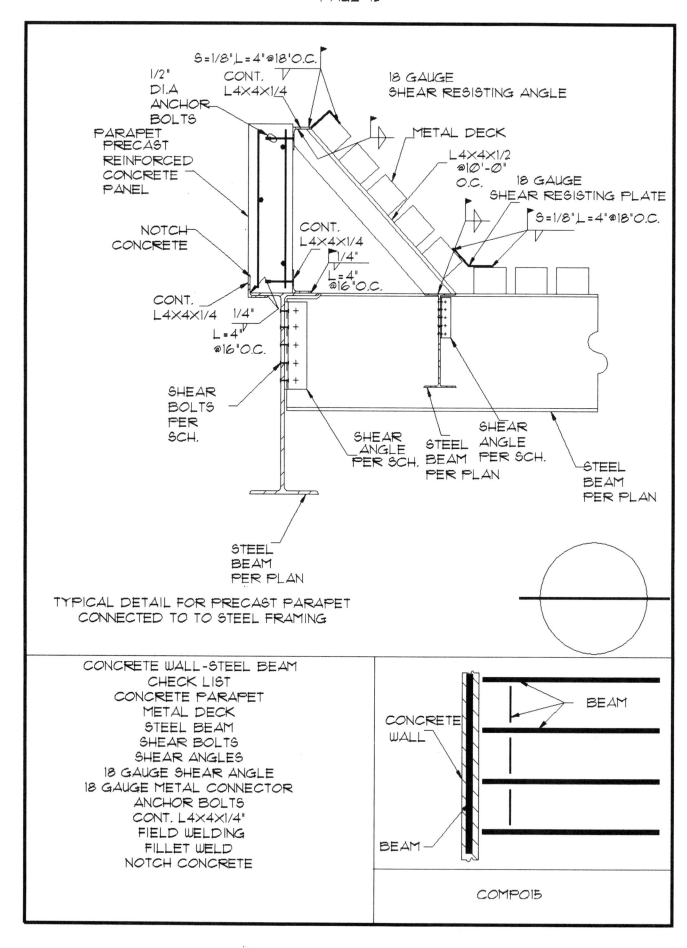

S=1/8" L=4" @18'O.C.
CONT. L4X4X1/4
18 GAUGE SHEAR RESISTING ANGLE
1/2" DI.A ANCHOR BOLTS
PARAPET PRECAST REINFORCED CONCRETE PANEL
METAL DECK
L4X4X1/2 @10'-0" O.C.
18 GAUGE SHEAR RESISTING PLATE
S=1/8" L=4" @18"O.C.
NOTCH CONCRETE
CONT. L4X4X1/4
1/4"
L=4" @16"O.C.
CONT. L4X4X1/4
1/4"
L=4" @16"O.C.
SHEAR BOLTS PER SCH.
SHEAR ANGLE PER SCH.
STEEL BEAM PER PLAN
SHEAR ANGLE PER SCH.
STEEL BEAM PER PLAN
STEEL BEAM PER PLAN

TYPICAL DETAIL FOR PRECAST PARAPET CONNECTED TO TO STEEL FRAMING

CONCRETE WALL-STEEL BEAM
CHECK LIST
CONCRETE PARAPET
METAL DECK
STEEL BEAM
SHEAR BOLTS
SHEAR ANGLES
18 GAUGE SHEAR ANGLE
18 GAUGE METAL CONNECTOR
ANCHOR BOLTS
CONT. L4X4X1/4"
FIELD WELDING
FILLET WELD
NOTCH CONCRETE

BEAM
CONCRETE WALL
BEAM

COMPO15

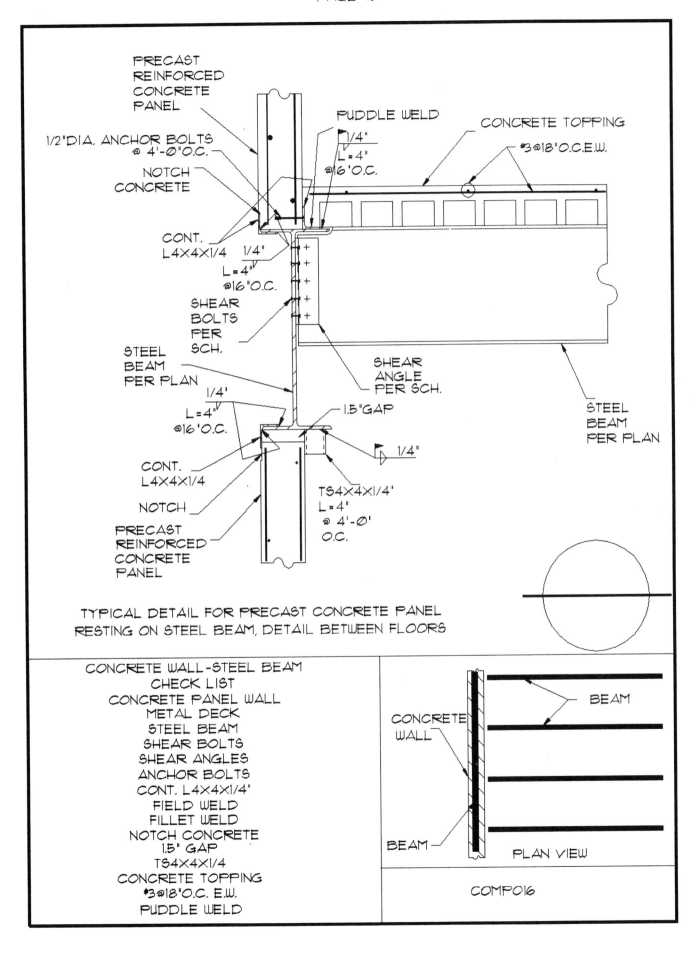

PRECAST
REINFORCED
CONCRETE
PANEL

PUDDLE WELD

CONCRETE TOPPING

1/2"DIA. ANCHOR BOLTS
@ 4'-0"O.C.

1/4"
L = 4"
@16'O.C.

#3@18"O.C.E.W.

NOTCH
CONCRETE

CONT.
L4X4X1/4

1/4"
L = 4"
@16"O.C.

SHEAR
BOLTS
PER
SCH.

SHEAR
ANGLE
PER SCH.

STEEL
BEAM
PER PLAN

1/4'
L=4"
@16'O.C.

1.5"GAP

1/4"

STEEL
BEAM
PER PLAN

CONT.
L4X4X1/4

NOTCH

PRECAST
REINFORCED
CONCRETE
PANEL

TS4X4X1/4'
L=4"
@ 4'-0'
O.C.

TYPICAL DETAIL FOR PRECAST CONCRETE PANEL
RESTING ON STEEL BEAM, DETAIL BETWEEN FLOORS

CONCRETE WALL-STEEL BEAM
CHECK LIST
CONCRETE PANEL WALL
METAL DECK
STEEL BEAM
SHEAR BOLTS
SHEAR ANGLES
ANCHOR BOLTS
CONT. L4X4X1/4"
FIELD WELD
FILLET WELD
NOTCH CONCRETE
1.5" GAP
TS4X4X1/4
CONCRETE TOPPING
#3@18"O.C. E.W.
PUDDLE WELD

BEAM

CONCRETE
WALL

BEAM

PLAN VIEW

COMP016

CHAPTER 5

METAL DECK

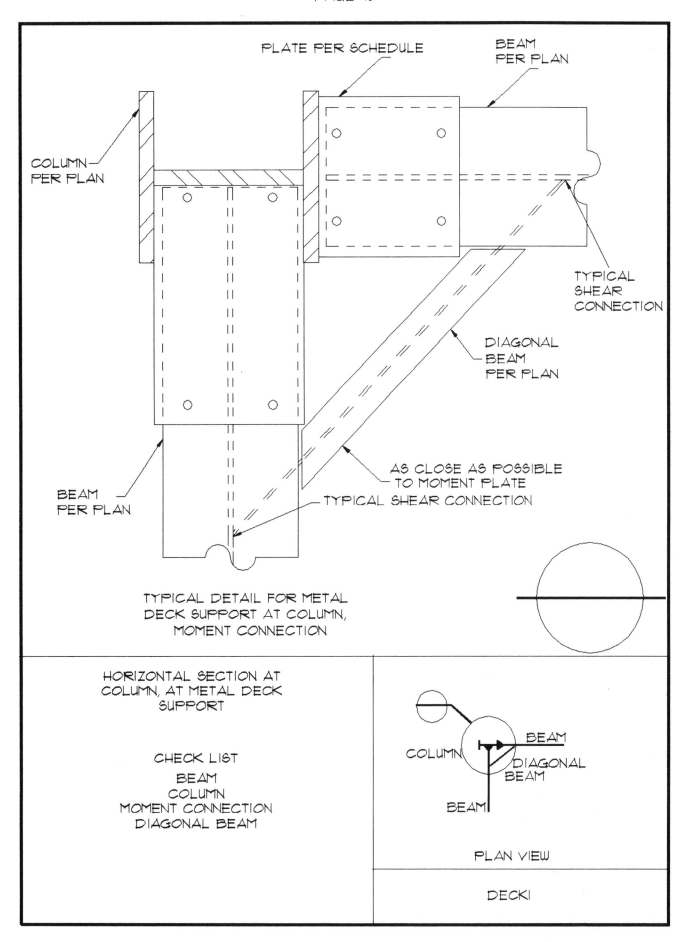

PLATE PER SCHEDULE

BEAM PER PLAN

COLUMN PER PLAN

TYPICAL SHEAR CONNECTION

DIAGONAL BEAM PER PLAN

BEAM PER PLAN

AS CLOSE AS POSSIBLE TO MOMENT PLATE

TYPICAL SHEAR CONNECTION

TYPICAL DETAIL FOR METAL DECK SUPPORT AT COLUMN, MOMENT CONNECTION

HORIZONTAL SECTION AT COLUMN, AT METAL DECK SUPPORT

CHECK LIST
BEAM
COLUMN
MOMENT CONNECTION
DIAGONAL BEAM

COLUMN

BEAM

DIAGONAL BEAM

BEAM

PLAN VIEW

DECK1

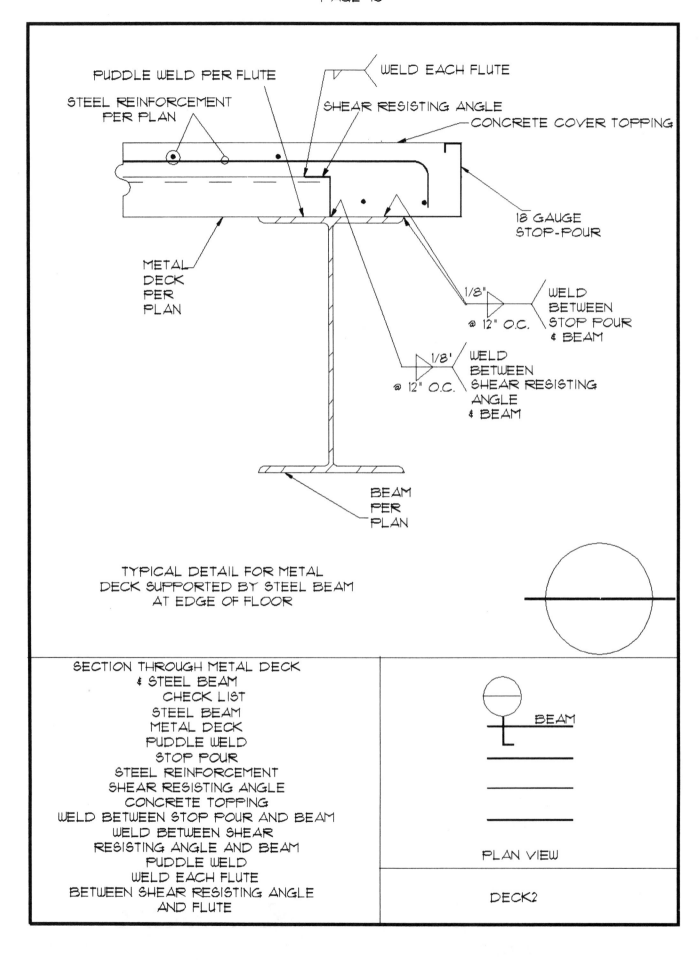

PUDDLE WELD PER FLUTE

WELD EACH FLUTE

STEEL REINFORCEMENT PER PLAN

SHEAR RESISTING ANGLE

CONCRETE COVER TOPPING

18 GAUGE STOP-POUR

METAL DECK PER PLAN

1/8" @ 12" O.C. WELD BETWEEN STOP POUR & BEAM

1/8' @ 12" O.C. WELD BETWEEN SHEAR RESISTING ANGLE & BEAM

BEAM PER PLAN

TYPICAL DETAIL FOR METAL DECK SUPPORTED BY STEEL BEAM AT EDGE OF FLOOR

SECTION THROUGH METAL DECK & STEEL BEAM
CHECK LIST
STEEL BEAM
METAL DECK
PUDDLE WELD
STOP POUR
STEEL REINFORCEMENT
SHEAR RESISTING ANGLE
CONCRETE TOPPING
WELD BETWEEN STOP POUR AND BEAM
WELD BETWEEN SHEAR
RESISTING ANGLE AND BEAM
PUDDLE WELD
WELD EACH FLUTE
BETWEEN SHEAR RESISTING ANGLE
AND FLUTE

BEAM

PLAN VIEW

DECK2

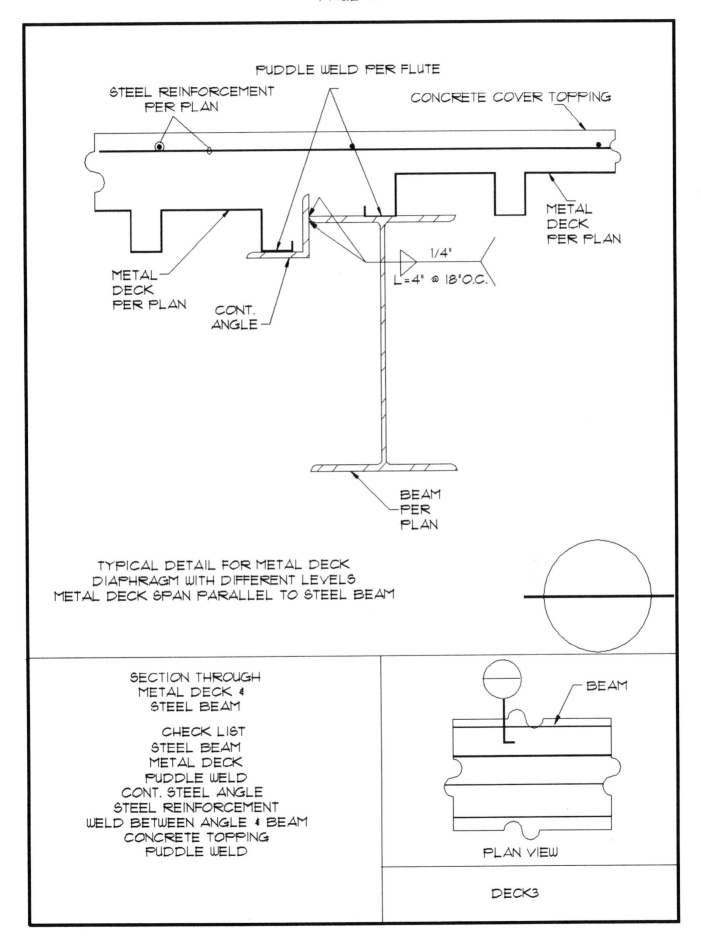

PUDDLE WELD PER FLUTE

STEEL REINFORCEMENT
PER PLAN

CONCRETE COVER TOPPING

METAL
DECK
PER PLAN

METAL
DECK
PER PLAN

CONT.
ANGLE

1/4"
L=4" @ 18"O.C.

BEAM
PER
PLAN

TYPICAL DETAIL FOR METAL DECK
DIAPHRAGM WITH DIFFERENT LEVELS
METAL DECK SPAN PARALLEL TO STEEL BEAM

SECTION THROUGH
METAL DECK &
STEEL BEAM

CHECK LIST
STEEL BEAM
METAL DECK
PUDDLE WELD
CONT. STEEL ANGLE
STEEL REINFORCEMENT
WELD BETWEEN ANGLE & BEAM
CONCRETE TOPPING
PUDDLE WELD

BEAM

PLAN VIEW

DECK3

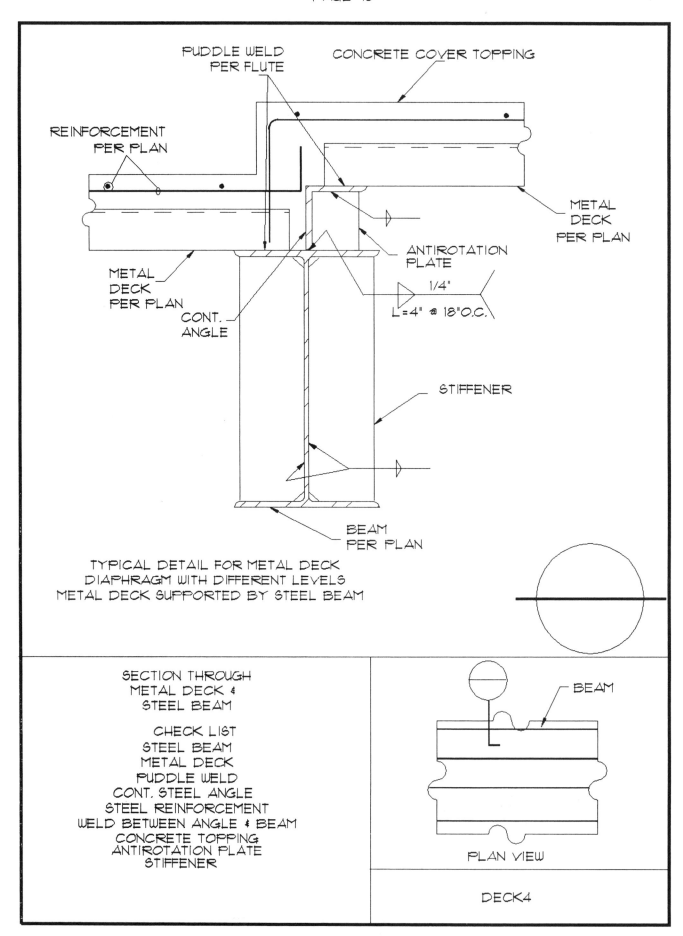

PUDDLE WELD
PER FLUTE

CONCRETE COVER TOPPING

REINFORCEMENT
PER PLAN

METAL
DECK
PER PLAN

ANTIROTATION
PLATE

METAL
DECK
PER PLAN

CONT.
ANGLE

1/4"
L=4" @ 18"O.C.

STIFFENER

BEAM
PER PLAN

TYPICAL DETAIL FOR METAL DECK
DIAPHRAGM WITH DIFFERENT LEVELS
METAL DECK SUPPORTED BY STEEL BEAM

SECTION THROUGH
METAL DECK &
STEEL BEAM

CHECK LIST
STEEL BEAM
METAL DECK
PUDDLE WELD
CONT. STEEL ANGLE
STEEL REINFORCEMENT
WELD BETWEEN ANGLE & BEAM
CONCRETE TOPPING
ANTIROTATION PLATE
STIFFENER

BEAM

PLAN VIEW

DECK4

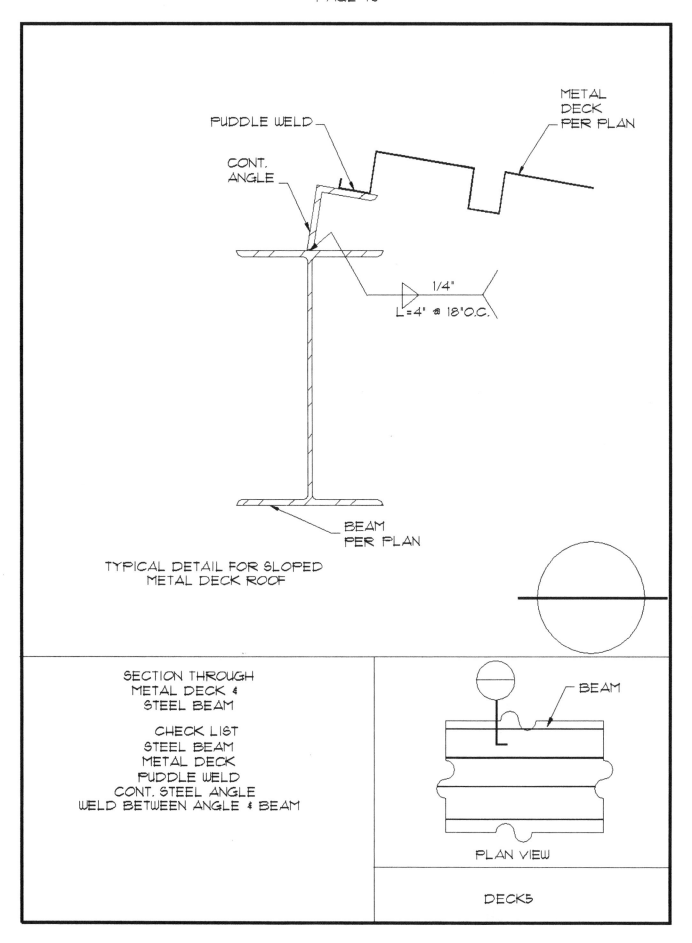

PUDDLE WELD

METAL DECK PER PLAN

CONT. ANGLE

1/4"

L=4" @ 18"O.C.

BEAM PER PLAN

TYPICAL DETAIL FOR SLOPED METAL DECK ROOF

SECTION THROUGH
METAL DECK &
STEEL BEAM

CHECK LIST
STEEL BEAM
METAL DECK
PUDDLE WELD
CONT. STEEL ANGLE
WELD BETWEEN ANGLE & BEAM

BEAM

PLAN VIEW

DECK5

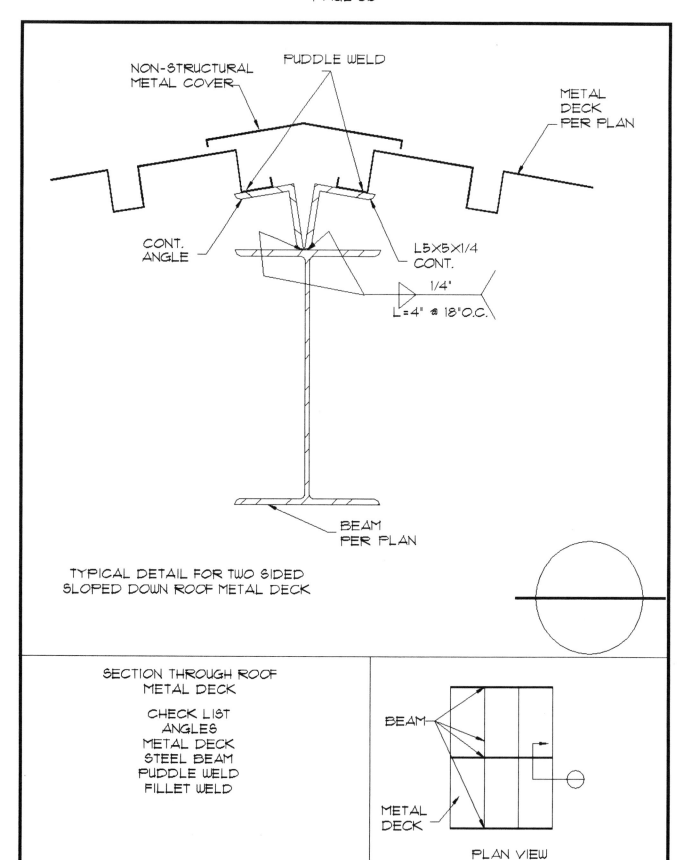

NON-STRUCTURAL
METAL COVER

PUDDLE WELD

METAL
DECK
PER PLAN

CONT.
ANGLE

L5X5X1/4
CONT.

1/4"

L=4" @ 18"O.C.

BEAM
PER PLAN

TYPICAL DETAIL FOR TWO SIDED
SLOPED DOWN ROOF METAL DECK

SECTION THROUGH ROOF
METAL DECK

CHECK LIST
ANGLES
METAL DECK
STEEL BEAM
PUDDLE WELD
FILLET WELD

BEAM

METAL
DECK

PLAN VIEW

DECK6

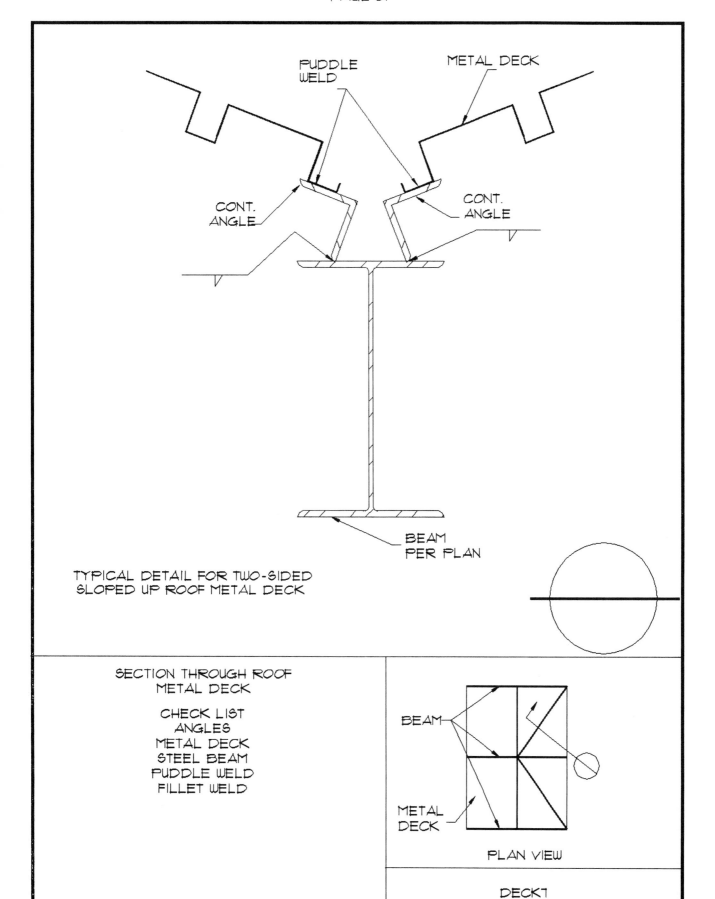

PUDDLE
WELD

METAL DECK

CONT.
ANGLE

CONT.
ANGLE

BEAM
PER PLAN

TYPICAL DETAIL FOR TWO-SIDED
SLOPED UP ROOF METAL DECK

SECTION THROUGH ROOF
METAL DECK

CHECK LIST
ANGLES
METAL DECK
STEEL BEAM
PUDDLE WELD
FILLET WELD

BEAM

METAL
DECK

PLAN VIEW

DECK1

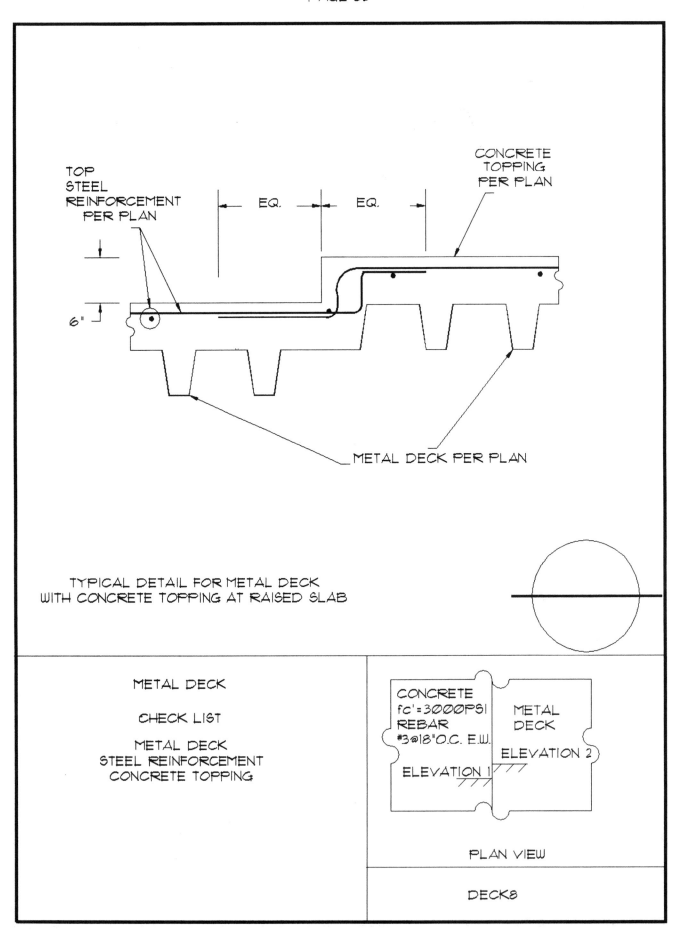

TOP STEEL REINFORCEMENT PER PLAN

CONCRETE TOPPING PER PLAN

EQ.

EQ.

6"

METAL DECK PER PLAN

TYPICAL DETAIL FOR METAL DECK
WITH CONCRETE TOPPING AT RAISED SLAB

METAL DECK

CHECK LIST

METAL DECK
STEEL REINFORCEMENT
CONCRETE TOPPING

CONCRETE
fc'=3000PSI
REBAR
#3@18"O.C. E.W.

ELEVATION 1

METAL
DECK

ELEVATION 2

PLAN VIEW

DECK8

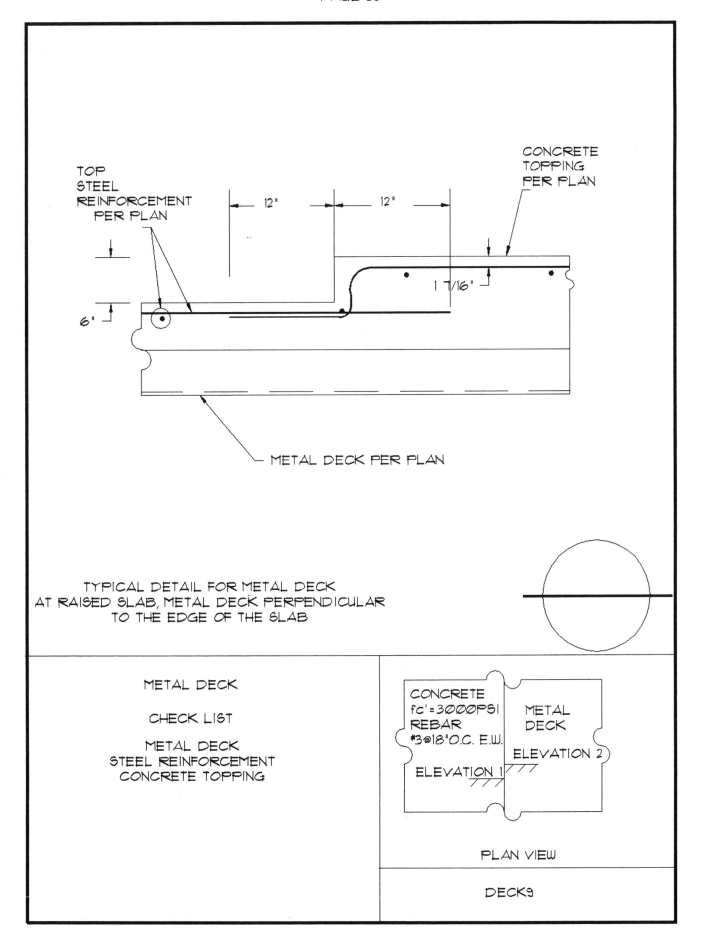

TOP STEEL REINFORCEMENT PER PLAN

CONCRETE TOPPING PER PLAN

12"

12"

1 7/16"

6"

METAL DECK PER PLAN

TYPICAL DETAIL FOR METAL DECK
AT RAISED SLAB, METAL DECK PERPENDICULAR
TO THE EDGE OF THE SLAB

METAL DECK

CHECK LIST

METAL DECK
STEEL REINFORCEMENT
CONCRETE TOPPING

CONCRETE
fc'=3000PSI
REBAR
#3@18"O.C. E.W.

METAL DECK

ELEVATION 1

ELEVATION 2

PLAN VIEW

DECK9

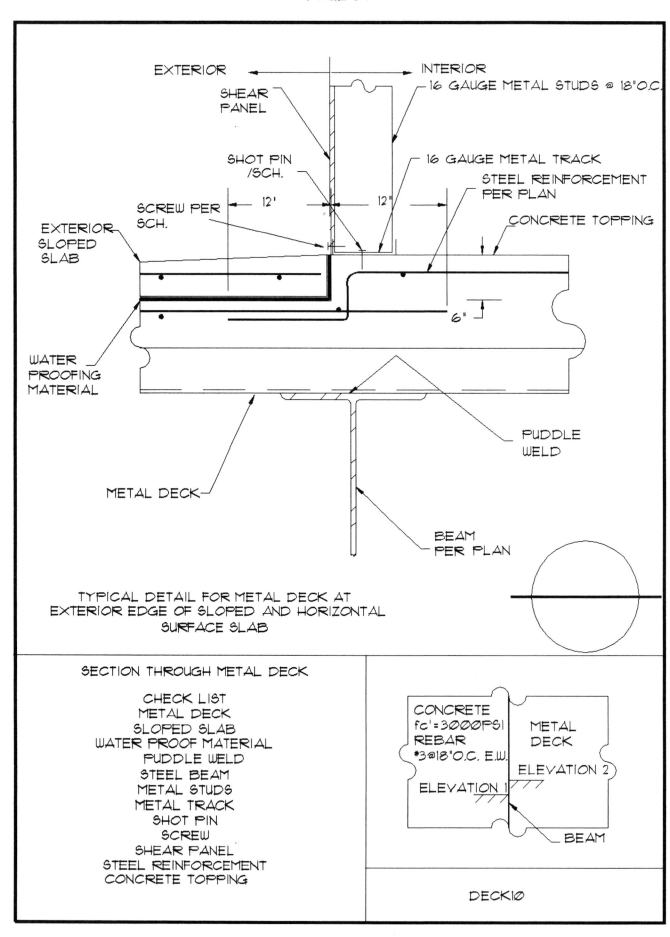

EXTERIOR ← → INTERIOR

SHEAR PANEL

16 GAUGE METAL STUDS @ 18"O.C.

SHOT PIN /SCH.

16 GAUGE METAL TRACK

STEEL REINFORCEMENT PER PLAN

SCREW PER SCH.

12'

12"

CONCRETE TOPPING

EXTERIOR SLOPED SLAB

WATER PROOFING MATERIAL

6"

PUDDLE WELD

METAL DECK

BEAM PER PLAN

TYPICAL DETAIL FOR METAL DECK AT
EXTERIOR EDGE OF SLOPED AND HORIZONTAL
SURFACE SLAB

SECTION THROUGH METAL DECK

CHECK LIST
METAL DECK
SLOPED SLAB
WATER PROOF MATERIAL
PUDDLE WELD
STEEL BEAM
METAL STUDS
METAL TRACK
SHOT PIN
SCREW
SHEAR PANEL
STEEL REINFORCEMENT
CONCRETE TOPPING

CONCRETE
fc'=3000PSI
REBAR
#3@18"O.C. E.W.

METAL DECK

ELEVATION 2

ELEVATION 1

BEAM

DECK10

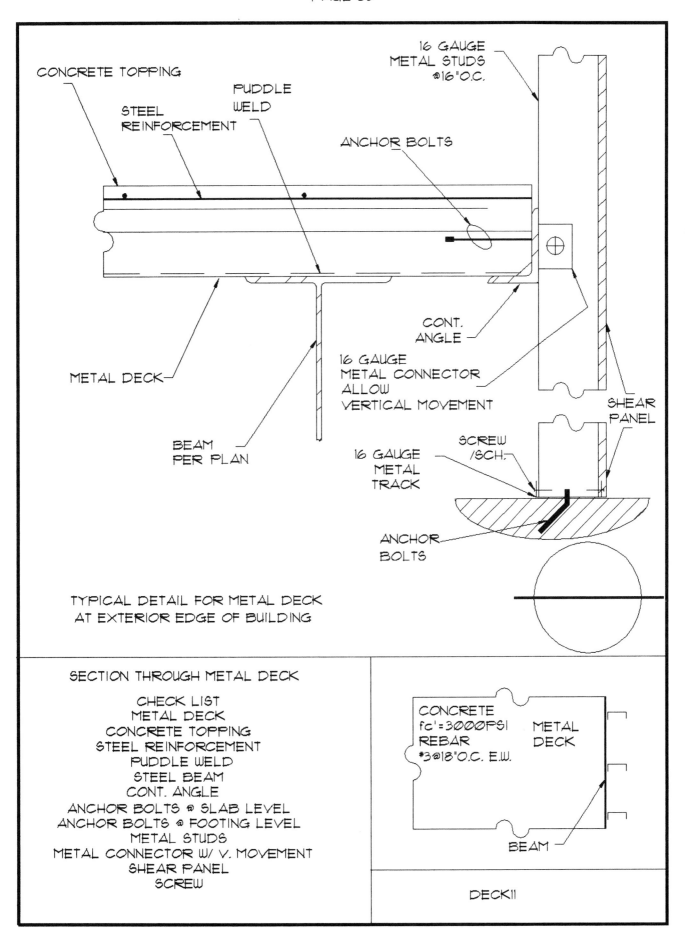

CONCRETE TOPPING

STEEL
REINFORCEMENT

PUDDLE
WELD

ANCHOR BOLTS

16 GAUGE
METAL STUDS
@16"O.C.

METAL DECK

BEAM
PER PLAN

CONT.
ANGLE

16 GAUGE
METAL CONNECTOR
ALLOW
VERTICAL MOVEMENT

SHEAR
PANEL

SCREW
/SCH.

16 GAUGE
METAL
TRACK

ANCHOR
BOLTS

TYPICAL DETAIL FOR METAL DECK
AT EXTERIOR EDGE OF BUILDING

SECTION THROUGH METAL DECK

CHECK LIST
METAL DECK
CONCRETE TOPPING
STEEL REINFORCEMENT
PUDDLE WELD
STEEL BEAM
CONT. ANGLE
ANCHOR BOLTS @ SLAB LEVEL
ANCHOR BOLTS @ FOOTING LEVEL
METAL STUDS
METAL CONNECTOR W/ V. MOVEMENT
SHEAR PANEL
SCREW

CONCRETE
fc'=3000PSI
REBAR
#3@18"O.C. E.W.

METAL
DECK

BEAM

DECK11

CHAPTER 6

FOUNDATION

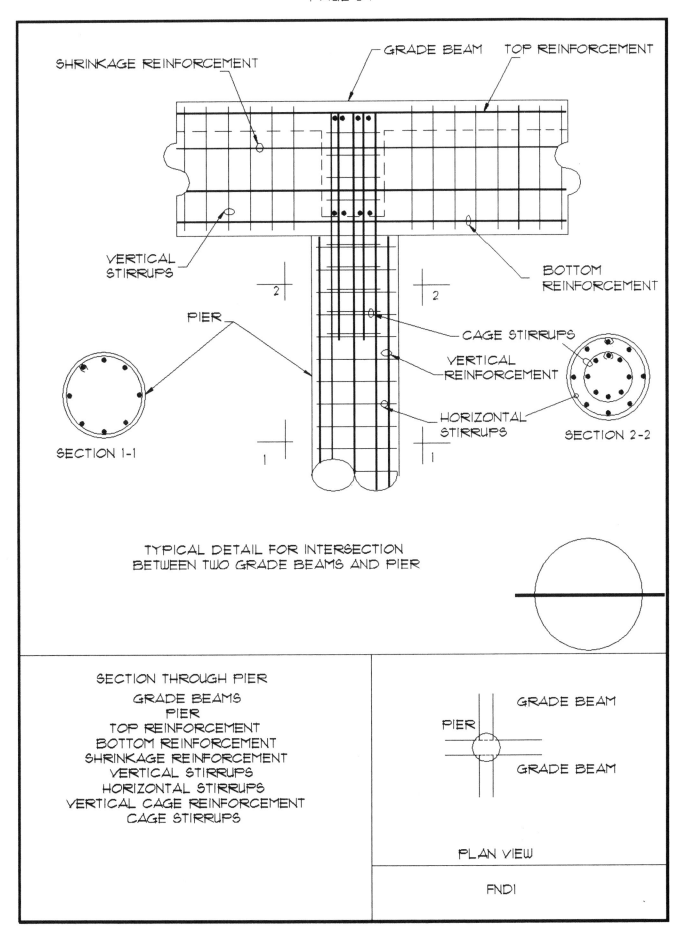

SHRINKAGE REINFORCEMENT

GRADE BEAM

TOP REINFORCEMENT

VERTICAL STIRRUPS

BOTTOM REINFORCEMENT

PIER

CAGE STIRRUPS

VERTICAL REINFORCEMENT

HORIZONTAL STIRRUPS

SECTION 1-1

SECTION 2-2

TYPICAL DETAIL FOR INTERSECTION
BETWEEN TWO GRADE BEAMS AND PIER

SECTION THROUGH PIER

GRADE BEAMS
PIER
TOP REINFORCEMENT
BOTTOM REINFORCEMENT
SHRINKAGE REINFORCEMENT
VERTICAL STIRRUPS
HORIZONTAL STIRRUPS
VERTICAL CAGE REINFORCEMENT
CAGE STIRRUPS

GRADE BEAM

PIER

GRADE BEAM

PLAN VIEW

FND1

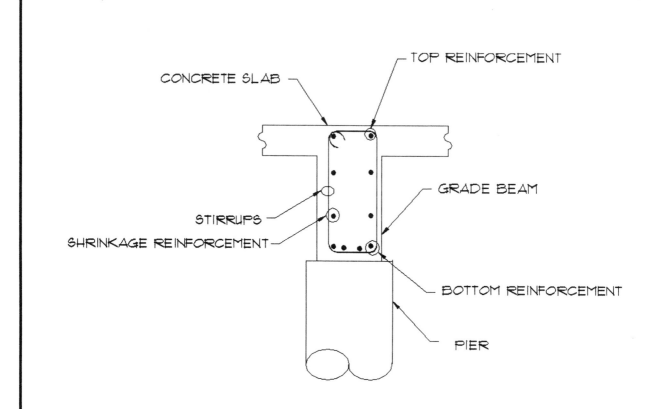

CONCRETE SLAB

TOP REINFORCEMENT

GRADE BEAM

STIRRUPS

SHRINKAGE REINFORCEMENT

BOTTOM REINFORCEMENT

PIER

TYPICAL DETAIL FOR GRADE BEAM ABOVE PIER

SECTION THROUGH GRADE BEAM

CHECK LIST
STIRRUPS
TOP REINFORCEMENT
BOTTOM REINFORCEMENT
SHRINKAGE REINFORCEMENT
GRADE BEAM
CONCRETE SLAB
PIER

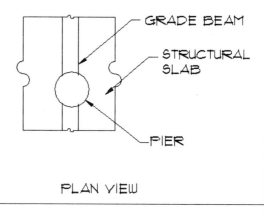

GRADE BEAM

STRUCTURAL SLAB

PIER

PLAN VIEW

FND2

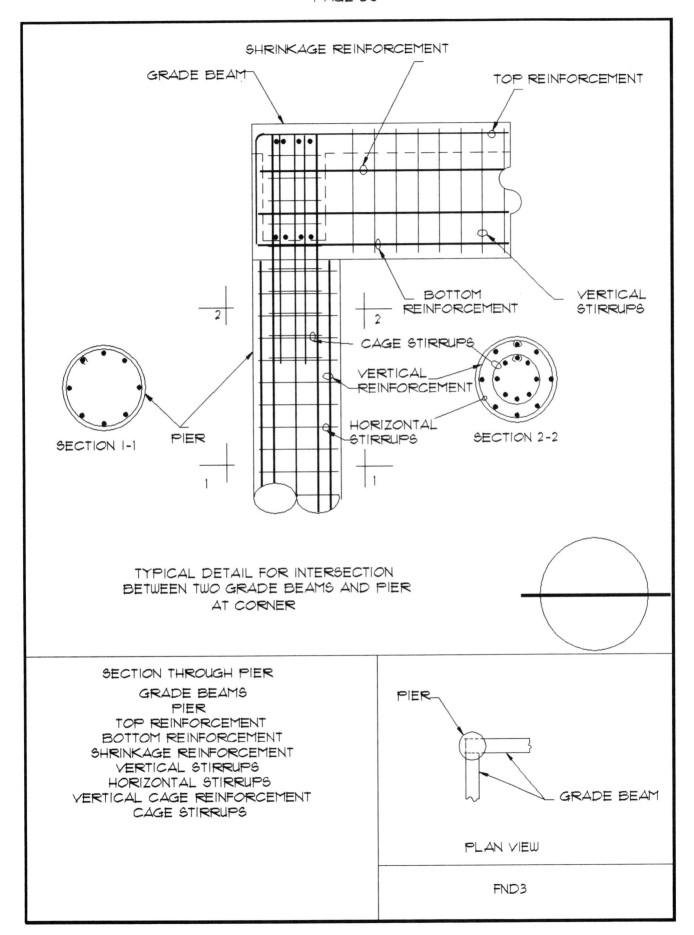

SHRINKAGE REINFORCEMENT

GRADE BEAM

TOP REINFORCEMENT

BOTTOM REINFORCEMENT

VERTICAL STIRRUPS

CAGE STIRRUPS

VERTICAL REINFORCEMENT

HORIZONTAL STIRRUPS

PIER

SECTION 1-1

SECTION 2-2

TYPICAL DETAIL FOR INTERSECTION
BETWEEN TWO GRADE BEAMS AND PIER
AT CORNER

SECTION THROUGH PIER

GRADE BEAMS
PIER
TOP REINFORCEMENT
BOTTOM REINFORCEMENT
SHRINKAGE REINFORCEMENT
VERTICAL STIRRUPS
HORIZONTAL STIRRUPS
VERTICAL CAGE REINFORCEMENT
CAGE STIRRUPS

PIER

GRADE BEAM

PLAN VIEW

FND3

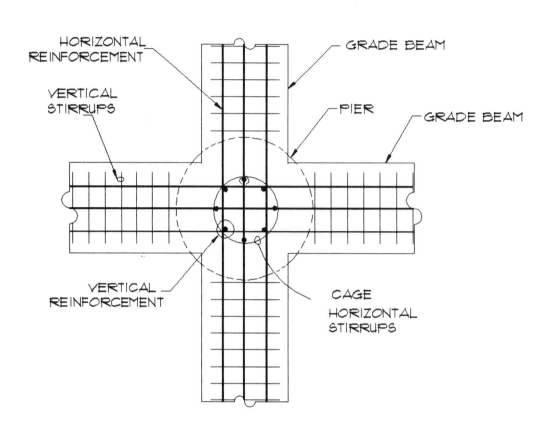

HORIZONTAL
REINFORCEMENT

VERTICAL
STIRRUPS

GRADE BEAM

PIER

GRADE BEAM

VERTICAL
REINFORCEMENT

CAGE
HORIZONTAL
STIRRUPS

TYPICAL DETAIL FOR INTERSECTION
BETWEEN TWO GRADE BEAMS, WITH
PIER , PLAN VIEW

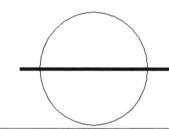

HORIZONTAL TOP VIEW
FOR GRADE BEAMS

CHECK LIST
GRADE BEAMS
HORIZONTAL REINFORCEMENT
VERTICAL STIRRUPS
HORIZONTAL STIRRUPS
VERTICAL CAGE REINFORCEMENT
PIER

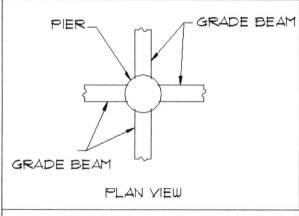

PIER

GRADE BEAM

GRADE BEAM

PLAN VIEW

FDN4

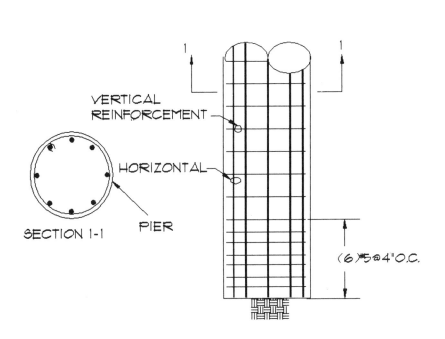

VERTICAL
REINFORCEMENT

HORIZONTAL

PIER

SECTION 1-1

(6)#5@4"O.C.

TYPICAL DETAIL FOR END PIER

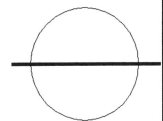

END PIER
CHECK LIST
PIER
VERTICAL REINFORCEMENT
HORIZONTAL REINFORCEMENT
HORIZONTAL STIRRUPS
(6)#5@4"O.C.

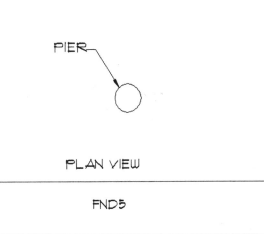

PIER

PLAN VIEW

FND5

PAGE 92

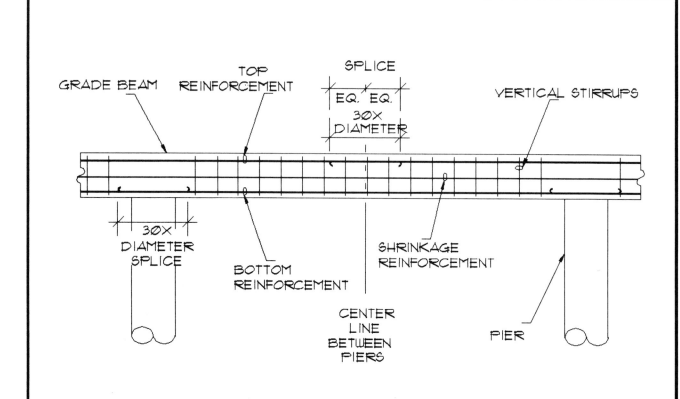

GRADE BEAM TOP REINFORCEMENT SPLICE VERTICAL STIRRUPS

EQ. EQ.
30X DIAMETER

30X DIAMETER SPLICE

BOTTOM REINFORCEMENT

SHRINKAGE REINFORCEMENT

CENTER LINE BETWEEN PIERS

PIER

TYPICAL DETAIL FOR CONTINUOUS GRADE BEAM
ABOVE PIERS SHOWING THE SPLICE FOR
LONGITUDINAL REINFORCEMENT

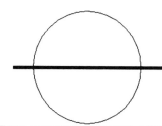

ELEVATION FOR GRADE BEAM

CHECK LIST
GRADE BEAM
PIER
SPLICES
TOP REINFORCEMENT
BOTTOM REINFORCEMENT
SHRINKAGE REINFORCEMENT
VERTICAL STIRRUPS

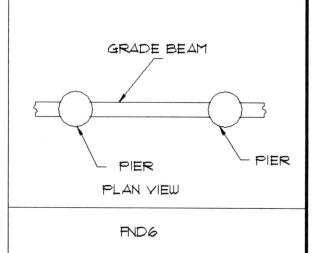

GRADE BEAM

PIER   PIER

PLAN VIEW

FND6

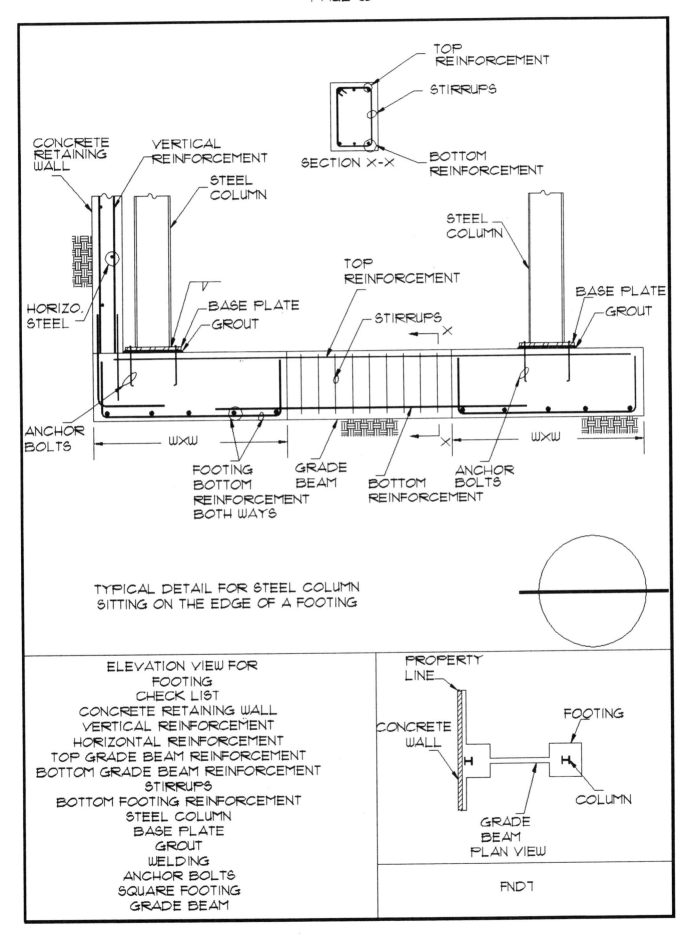

TOP
REINFORCEMENT

STIRRUPS

SECTION X-X

BOTTOM
REINFORCEMENT

CONCRETE
RETAINING
WALL

VERTICAL
REINFORCEMENT

STEEL
COLUMN

STEEL
COLUMN

TOP
REINFORCEMENT

BASE PLATE

GROUT

BASE PLATE

GROUT

HORIZO.
STEEL

STIRRUPS

ANCHOR
BOLTS

WXW

FOOTING
BOTTOM
REINFORCEMENT
BOTH WAYS

GRADE
BEAM

BOTTOM
REINFORCEMENT

ANCHOR
BOLTS

WXW

TYPICAL DETAIL FOR STEEL COLUMN
SITTING ON THE EDGE OF A FOOTING

ELEVATION VIEW FOR
FOOTING
CHECK LIST
CONCRETE RETAINING WALL
VERTICAL REINFORCEMENT
HORIZONTAL REINFORCEMENT
TOP GRADE BEAM REINFORCEMENT
BOTTOM GRADE BEAM REINFORCEMENT
STIRRUPS
BOTTOM FOOTING REINFORCEMENT
STEEL COLUMN
BASE PLATE
GROUT
WELDING
ANCHOR BOLTS
SQUARE FOOTING
GRADE BEAM

PROPERTY
LINE

CONCRETE
WALL

FOOTING

GRADE
BEAM
PLAN VIEW

COLUMN

FND'T

CHAPTER 7

STEEL MOMENT CONNECTION

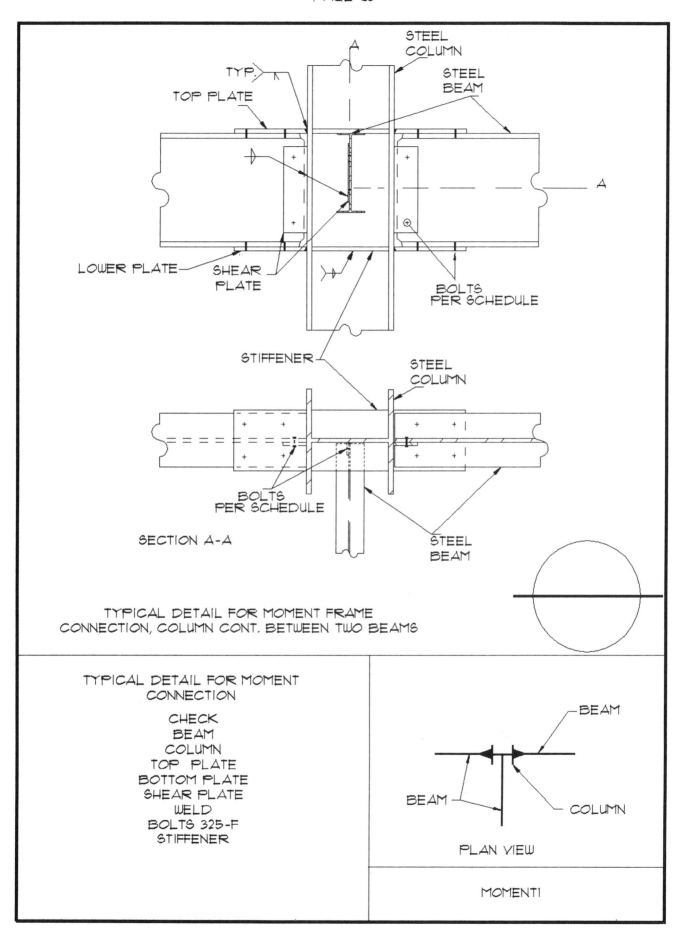

STEEL
COLUMN

STEEL
BEAM

TYP.

TOP PLATE

A

A

LOWER PLATE

SHEAR
PLATE

BOLTS
PER SCHEDULE

STIFFENER

STEEL
COLUMN

BOLTS
PER SCHEDULE

SECTION A-A

STEEL
BEAM

TYPICAL DETAIL FOR MOMENT FRAME
CONNECTION, COLUMN CONT. BETWEEN TWO BEAMS

TYPICAL DETAIL FOR MOMENT
CONNECTION

CHECK
BEAM
COLUMN
TOP PLATE
BOTTOM PLATE
SHEAR PLATE
WELD
BOLTS 325-F
STIFFENER

BEAM

BEAM

COLUMN

PLAN VIEW

MOMENT1

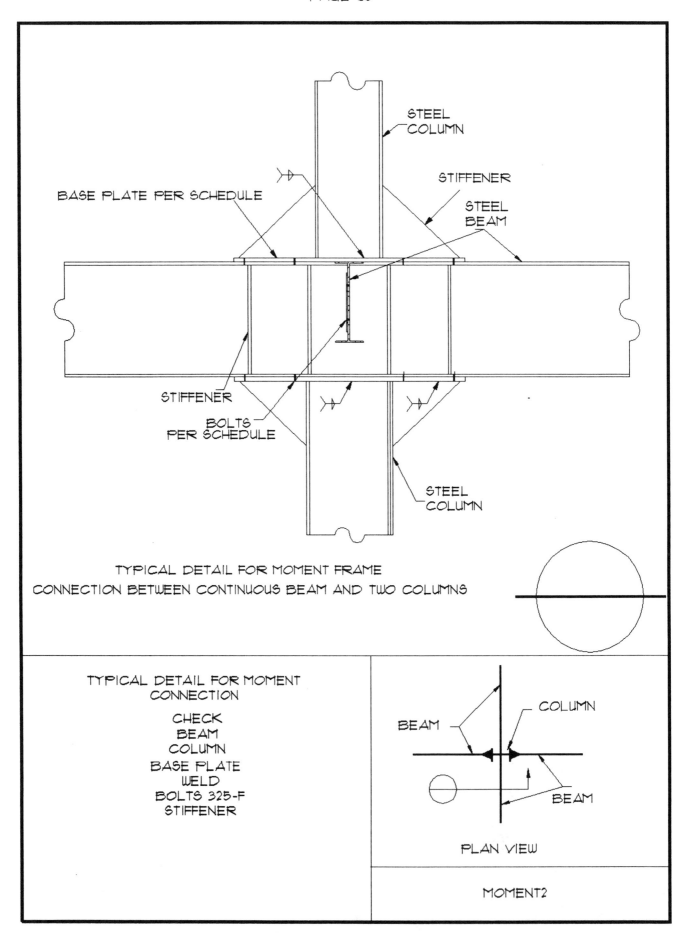

STEEL COLUMN

STIFFENER

STEEL BEAM

BASE PLATE PER SCHEDULE

STIFFENER

BOLTS PER SCHEDULE

STEEL COLUMN

TYPICAL DETAIL FOR MOMENT FRAME
CONNECTION BETWEEN CONTINUOUS BEAM AND TWO COLUMNS

TYPICAL DETAIL FOR MOMENT
CONNECTION

CHECK
BEAM
COLUMN
BASE PLATE
WELD
BOLTS 325-F
STIFFENER

COLUMN

BEAM

BEAM

PLAN VIEW

MOMENT2

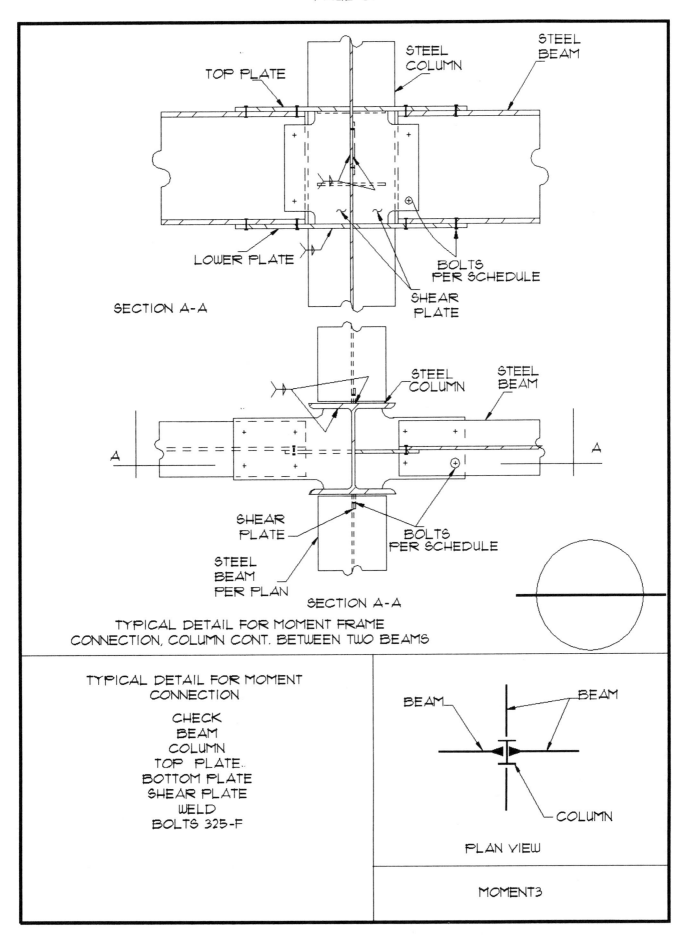

STEEL
COLUMN

STEEL
BEAM

TOP PLATE

LOWER PLATE

BOLTS
PER SCHEDULE

SHEAR
PLATE

SECTION A-A

STEEL
COLUMN

STEEL
BEAM

A   A

SHEAR
PLATE

BOLTS
PER SCHEDULE

STEEL
BEAM
PER PLAN

SECTION A-A

TYPICAL DETAIL FOR MOMENT FRAME
CONNECTION, COLUMN CONT. BETWEEN TWO BEAMS

TYPICAL DETAIL FOR MOMENT
CONNECTION

CHECK
BEAM
COLUMN
TOP PLATE
BOTTOM PLATE
SHEAR PLATE
WELD
BOLTS 325-F

BEAM   BEAM

COLUMN

PLAN VIEW

MOMENT3

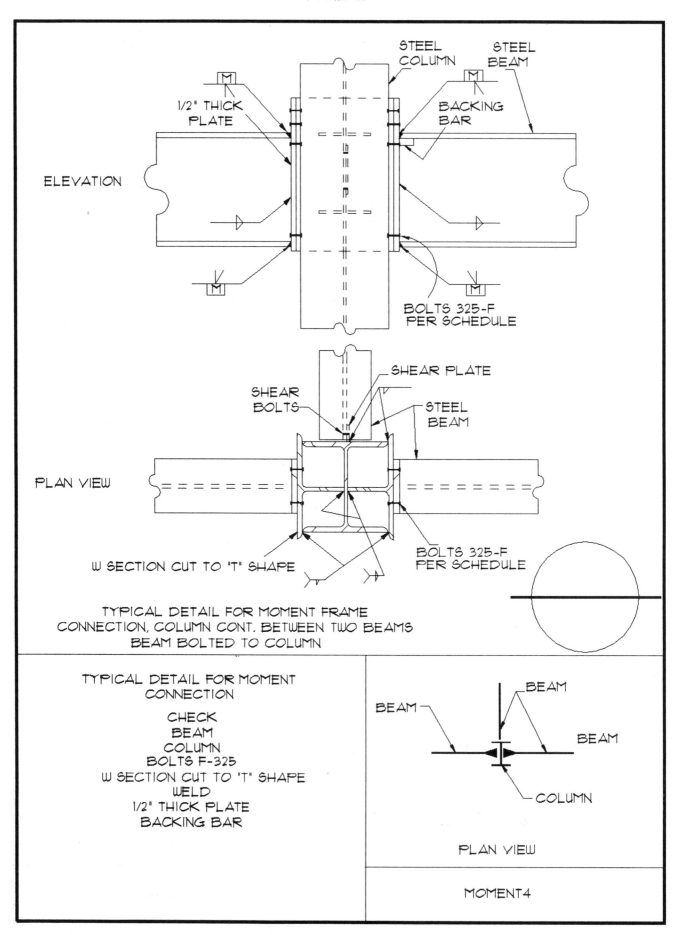

STEEL COLUMN

STEEL BEAM

1/2" THICK PLATE

BACKING BAR

ELEVATION

BOLTS 325-F PER SCHEDULE

SHEAR PLATE

SHEAR BOLTS

STEEL BEAM

PLAN VIEW

W SECTION CUT TO "T" SHAPE

BOLTS 325-F PER SCHEDULE

TYPICAL DETAIL FOR MOMENT FRAME
CONNECTION, COLUMN CONT. BETWEEN TWO BEAMS
BEAM BOLTED TO COLUMN

TYPICAL DETAIL FOR MOMENT
CONNECTION

CHECK
BEAM
COLUMN
BOLTS F-325
W SECTION CUT TO 'T" SHAPE
WELD
1/2" THICK PLATE
BACKING BAR

BEAM

BEAM

BEAM

COLUMN

PLAN VIEW

MOMENT4

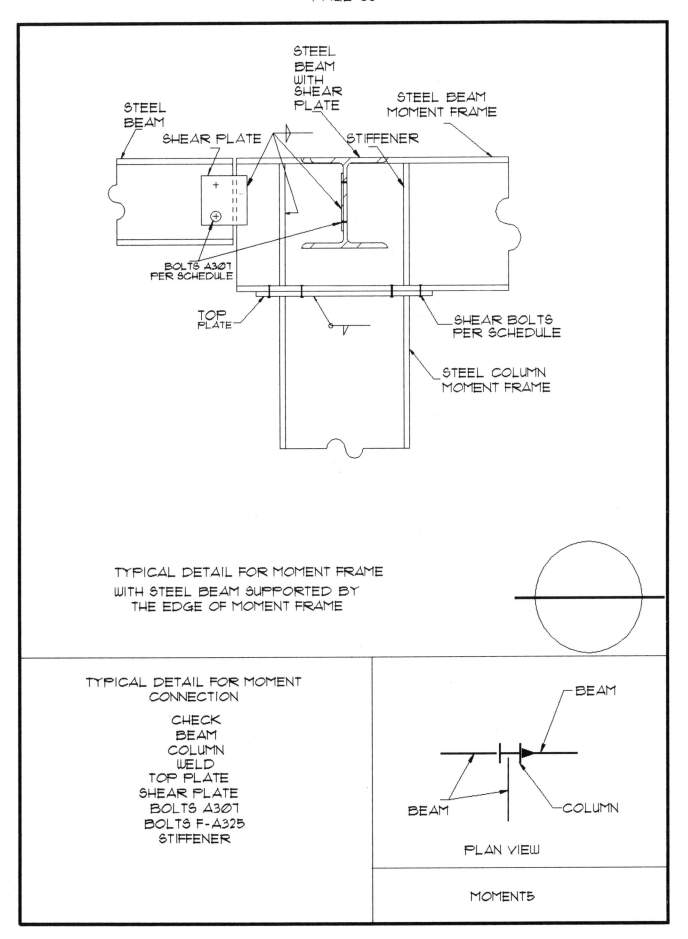

STEEL
BEAM
WITH
SHEAR
PLATE

STEEL
BEAM

SHEAR PLATE

STEEL BEAM
MOMENT FRAME

STIFFENER

BOLTS A307
PER SCHEDULE

TOP
PLATE

SHEAR BOLTS
PER SCHEDULE

STEEL COLUMN
MOMENT FRAME

TYPICAL DETAIL FOR MOMENT FRAME
WITH STEEL BEAM SUPPORTED BY
THE EDGE OF MOMENT FRAME

TYPICAL DETAIL FOR MOMENT
CONNECTION

CHECK
BEAM
COLUMN
WELD
TOP PLATE
SHEAR PLATE
BOLTS A307
BOLTS F-A325
STIFFENER

BEAM

BEAM

COLUMN

PLAN VIEW

MOMENT5

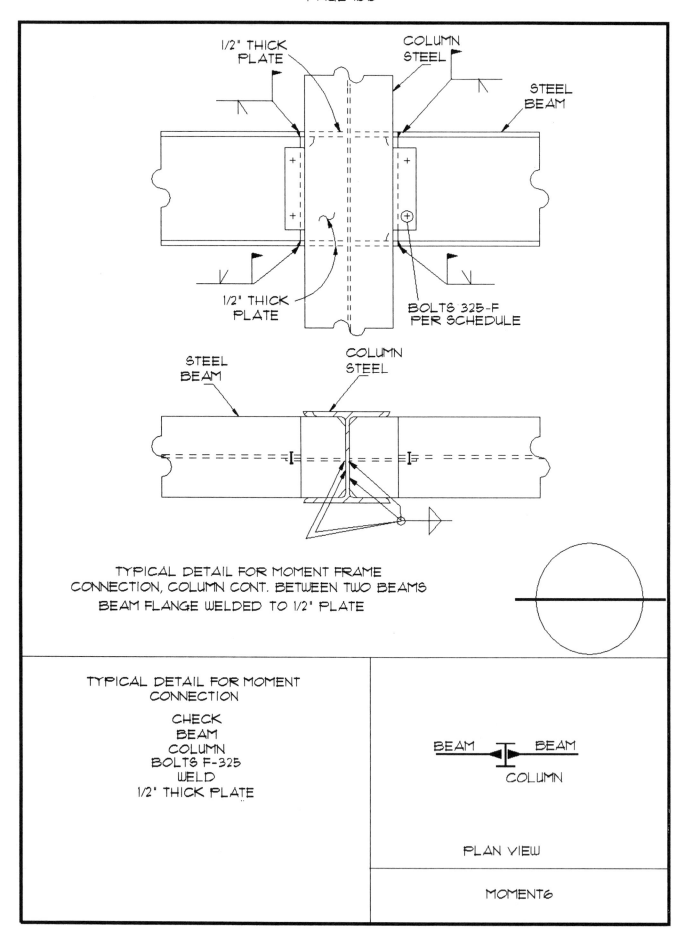

1/2" THICK PLATE

COLUMN STEEL

STEEL BEAM

1/2" THICK PLATE

BOLTS 325-F PER SCHEDULE

STEEL BEAM

COLUMN STEEL

TYPICAL DETAIL FOR MOMENT FRAME
CONNECTION, COLUMN CONT. BETWEEN TWO BEAMS
BEAM FLANGE WELDED TO 1/2" PLATE

TYPICAL DETAIL FOR MOMENT
CONNECTION

CHECK
BEAM
COLUMN
BOLTS F-325
WELD
1/2" THICK PLATE

BEAM   BEAM
COLUMN

PLAN VIEW

MOMENT6

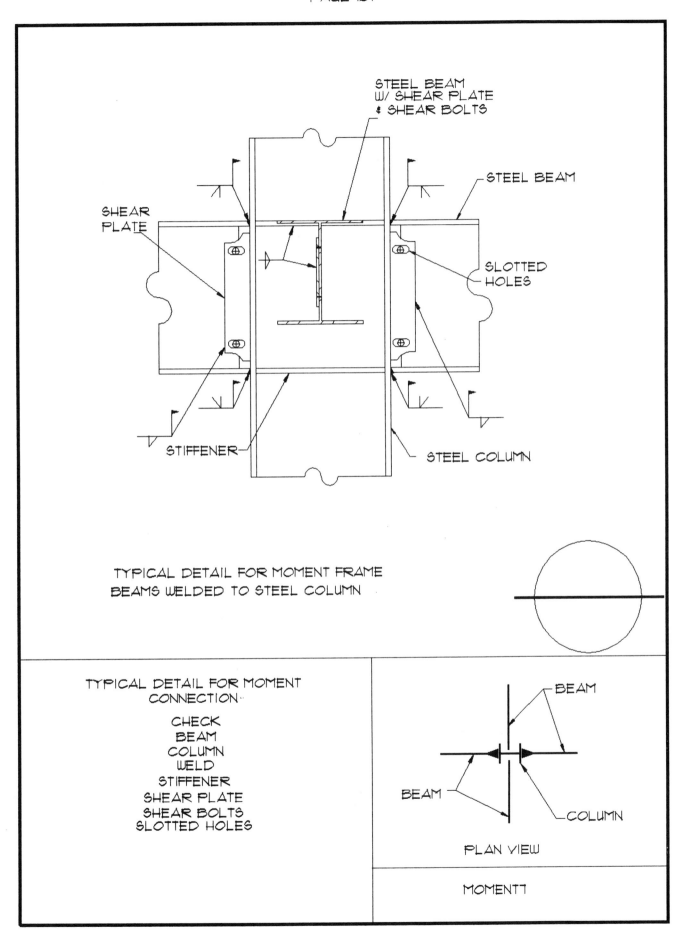

STEEL BEAM
W/ SHEAR PLATE
& SHEAR BOLTS

STEEL BEAM

SHEAR
PLATE

SLOTTED
HOLES

STIFFENER

STEEL COLUMN

TYPICAL DETAIL FOR MOMENT FRAME
BEAMS WELDED TO STEEL COLUMN

TYPICAL DETAIL FOR MOMENT
CONNECTION

CHECK
BEAM
COLUMN
WELD
STIFFENER
SHEAR PLATE
SHEAR BOLTS
SLOTTED HOLES

BEAM

BEAM

COLUMN

PLAN VIEW

MOMENT7

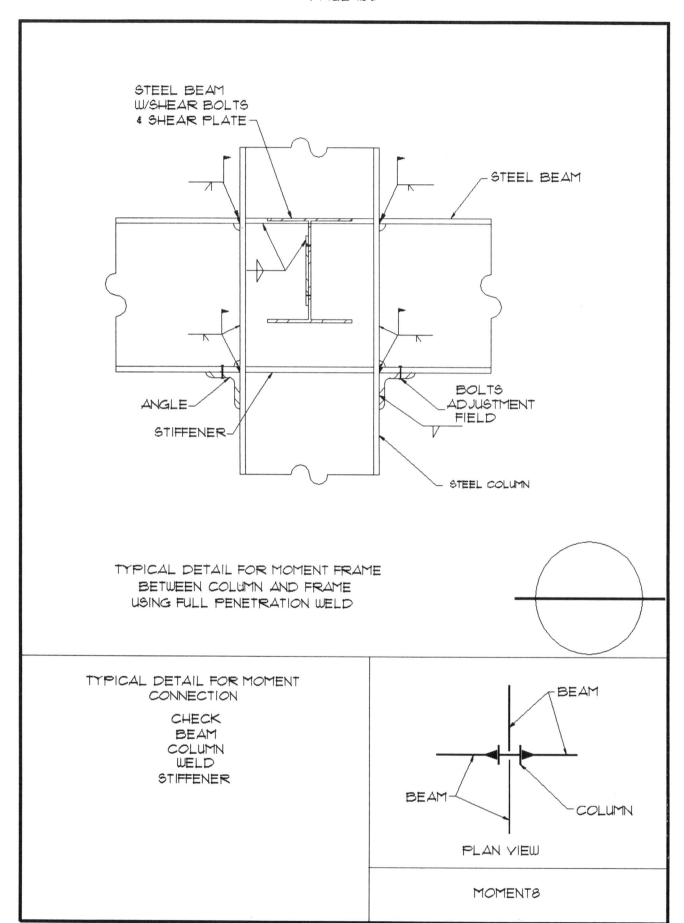

STEEL BEAM
W/SHEAR BOLTS
& SHEAR PLATE

STEEL BEAM

ANGLE

STIFFENER

BOLTS
ADJUSTMENT
FIELD

STEEL COLUMN

TYPICAL DETAIL FOR MOMENT FRAME
BETWEEN COLUMN AND FRAME
USING FULL PENETRATION WELD

TYPICAL DETAIL FOR MOMENT
CONNECTION

CHECK
BEAM
COLUMN
WELD
STIFFENER

BEAM

BEAM

COLUMN

PLAN VIEW

MOMENT8

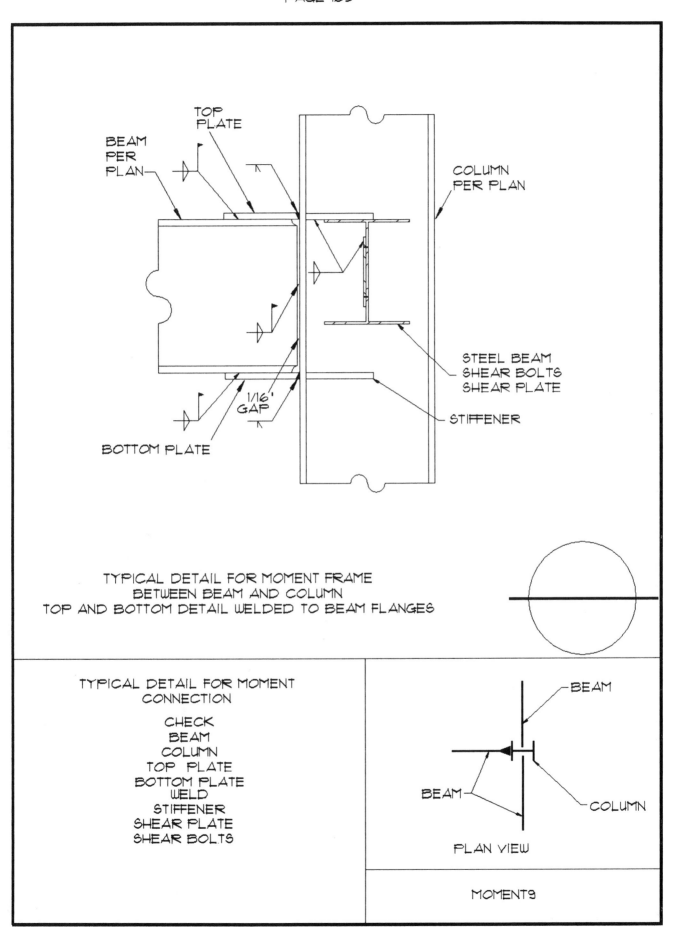

BEAM
PER
PLAN

TOP
PLATE

COLUMN
PER PLAN

STEEL BEAM
SHEAR BOLTS
SHEAR PLATE

STIFFENER

1/16'
GAP

BOTTOM PLATE

TYPICAL DETAIL FOR MOMENT FRAME
BETWEEN BEAM AND COLUMN
TOP AND BOTTOM DETAIL WELDED TO BEAM FLANGES

TYPICAL DETAIL FOR MOMENT
CONNECTION

CHECK
BEAM
COLUMN
TOP PLATE
BOTTOM PLATE
WELD
STIFFENER
SHEAR PLATE
SHEAR BOLTS

BEAM

BEAM

COLUMN

PLAN VIEW

MOMENT9

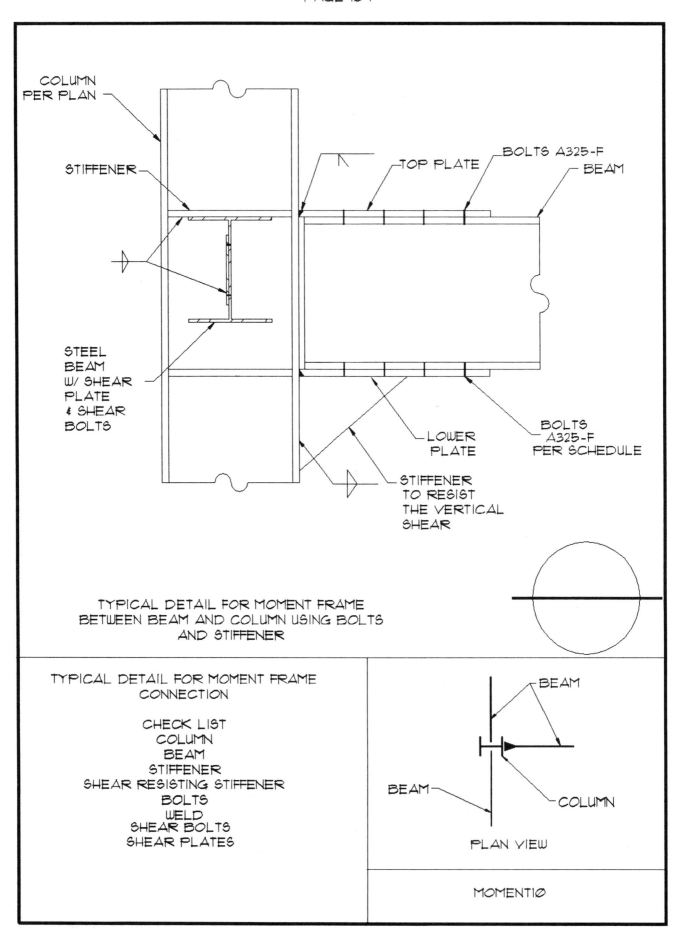

COLUMN
PER PLAN

STIFFENER

BOLTS A325-F

TOP PLATE

BEAM

STEEL
BEAM
W/ SHEAR
PLATE
& SHEAR
BOLTS

LOWER
PLATE

BOLTS
A325-F
PER SCHEDULE

STIFFENER
TO RESIST
THE VERTICAL
SHEAR

TYPICAL DETAIL FOR MOMENT FRAME
BETWEEN BEAM AND COLUMN USING BOLTS
AND STIFFENER

TYPICAL DETAIL FOR MOMENT FRAME
CONNECTION

CHECK LIST
COLUMN
BEAM
STIFFENER
SHEAR RESISTING STIFFENER
BOLTS
WELD
SHEAR BOLTS
SHEAR PLATES

BEAM

BEAM

COLUMN

PLAN VIEW

MOMENT10

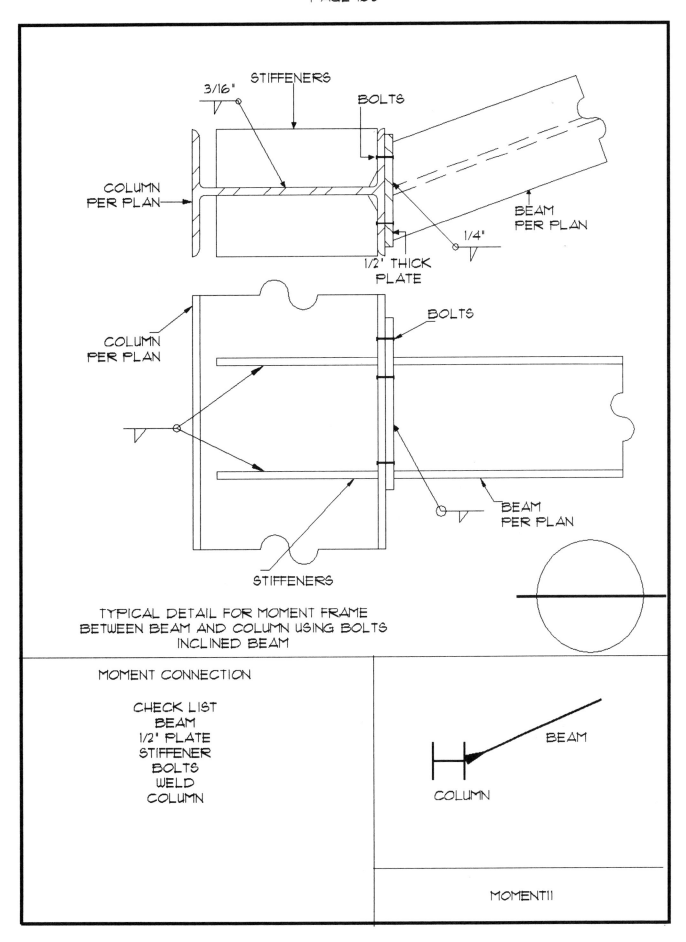

STIFFENERS

3/16"

BOLTS

COLUMN
PER PLAN

BEAM
PER PLAN

1/4"

1/2" THICK
PLATE

COLUMN
PER PLAN

BOLTS

BEAM
PER PLAN

STIFFENERS

TYPICAL DETAIL FOR MOMENT FRAME
BETWEEN BEAM AND COLUMN USING BOLTS
INCLINED BEAM

MOMENT CONNECTION

CHECK LIST
BEAM
1/2" PLATE
STIFFENER
BOLTS
WELD
COLUMN

BEAM

COLUMN

MOMENT11

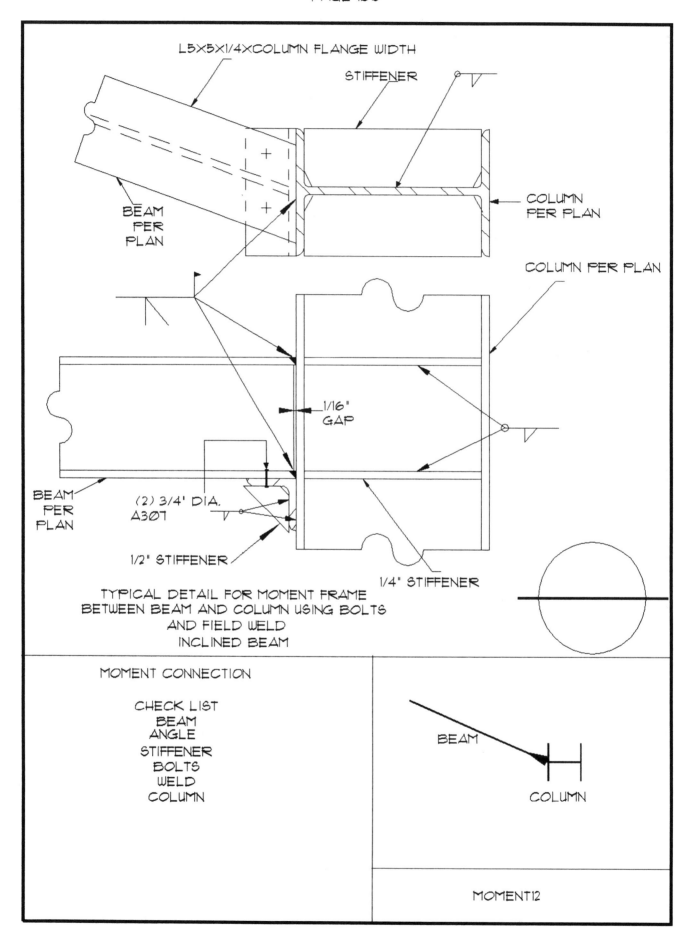

L5X5X1/4XCOLUMN FLANGE WIDTH

STIFFENER

COLUMN
PER PLAN

BEAM
PER
PLAN

COLUMN PER PLAN

1/16"
GAP

BEAM
PER
PLAN

(2) 3/4' DIA.
A307

1/2" STIFFENER

1/4" STIFFENER

TYPICAL DETAIL FOR MOMENT FRAME
BETWEEN BEAM AND COLUMN USING BOLTS
AND FIELD WELD
INCLINED BEAM

MOMENT CONNECTION

CHECK LIST
BEAM
ANGLE
STIFFENER
BOLTS
WELD
COLUMN

BEAM

COLUMN

MOMENT12

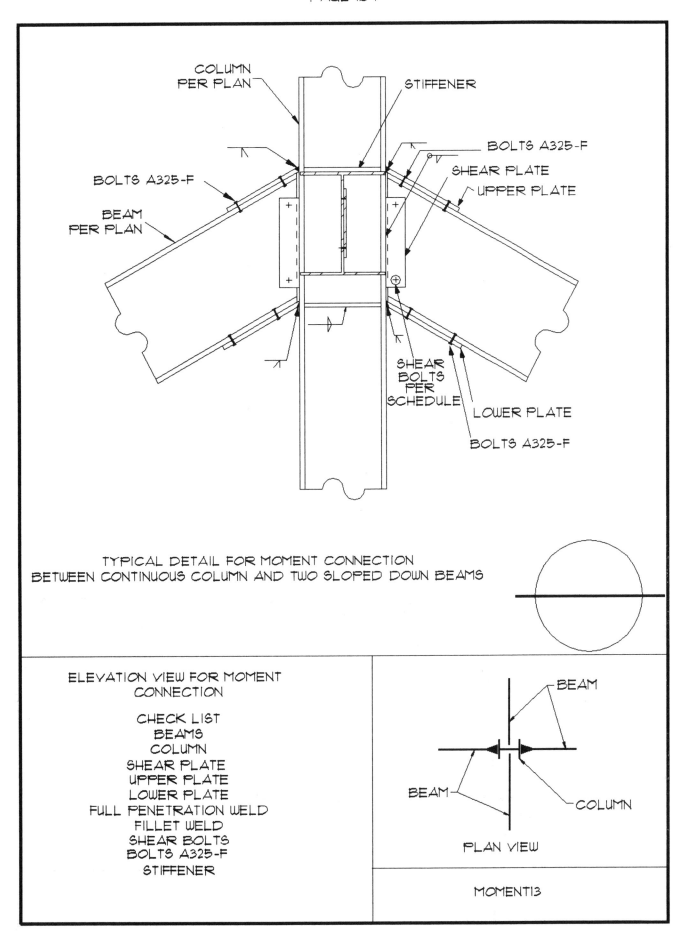

COLUMN
PER PLAN

STIFFENER

BOLTS A325-F

BOLTS A325-F

SHEAR PLATE

UPPER PLATE

BEAM
PER PLAN

SHEAR
BOLTS
PER
SCHEDULE

LOWER PLATE

BOLTS A325-F

TYPICAL DETAIL FOR MOMENT CONNECTION
BETWEEN CONTINUOUS COLUMN AND TWO SLOPED DOWN BEAMS

ELEVATION VIEW FOR MOMENT
CONNECTION

CHECK LIST
BEAMS
COLUMN
SHEAR PLATE
UPPER PLATE
LOWER PLATE
FULL PENETRATION WELD
FILLET WELD
SHEAR BOLTS
BOLTS A325-F
STIFFENER

BEAM

BEAM

COLUMN

PLAN VIEW

MOMENT13

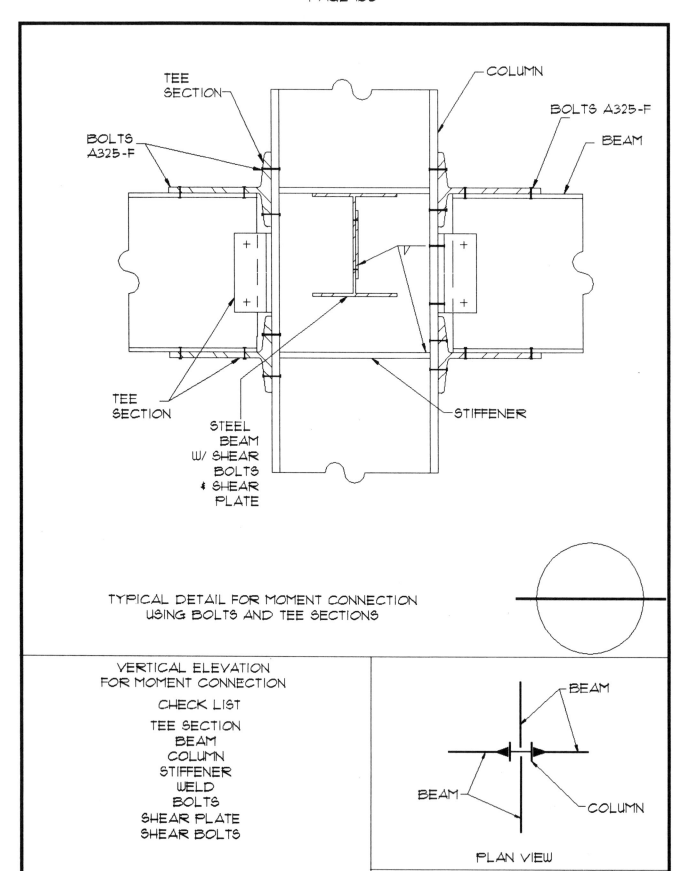

TEE SECTION

COLUMN

BOLTS A325-F

BOLTS A325-F

BEAM

TEE SECTION

STIFFENER

STEEL BEAM W/ SHEAR BOLTS & SHEAR PLATE

TYPICAL DETAIL FOR MOMENT CONNECTION USING BOLTS AND TEE SECTIONS

VERTICAL ELEVATION FOR MOMENT CONNECTION

CHECK LIST

TEE SECTION
BEAM
COLUMN
STIFFENER
WELD
BOLTS
SHEAR PLATE
SHEAR BOLTS

BEAM

BEAM

COLUMN

PLAN VIEW

MOMENT14

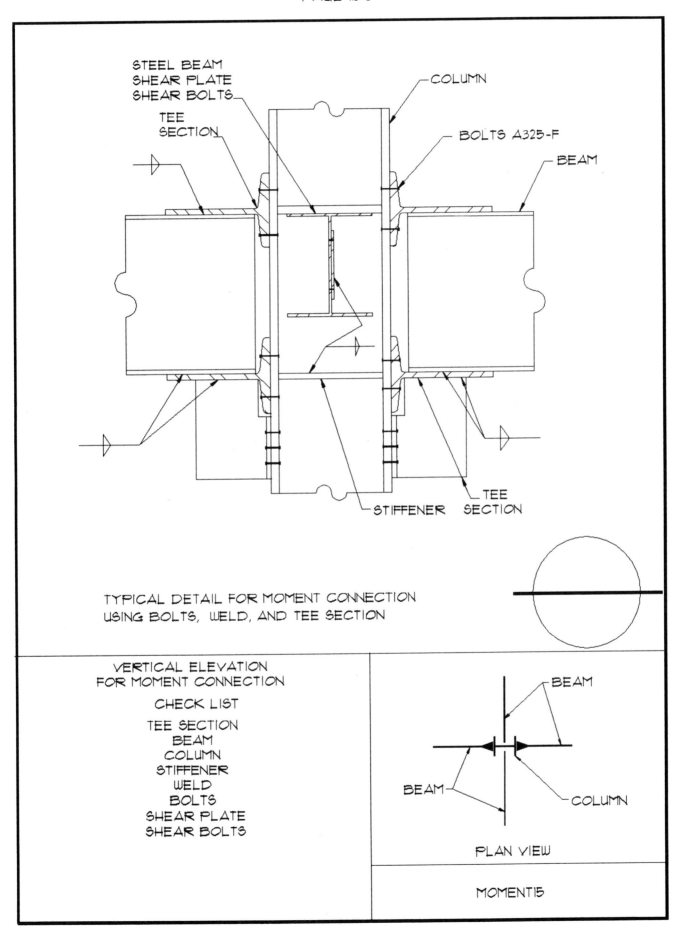

STEEL BEAM
SHEAR PLATE
SHEAR BOLTS

TEE
SECTION

COLUMN

BOLTS A325-F

BEAM

STIFFENER

TEE
SECTION

TYPICAL DETAIL FOR MOMENT CONNECTION
USING BOLTS, WELD, AND TEE SECTION

VERTICAL ELEVATION
FOR MOMENT CONNECTION

CHECK LIST

TEE SECTION
BEAM
COLUMN
STIFFENER
WELD
BOLTS
SHEAR PLATE
SHEAR BOLTS

BEAM

BEAM

COLUMN

PLAN VIEW

MOMENT15

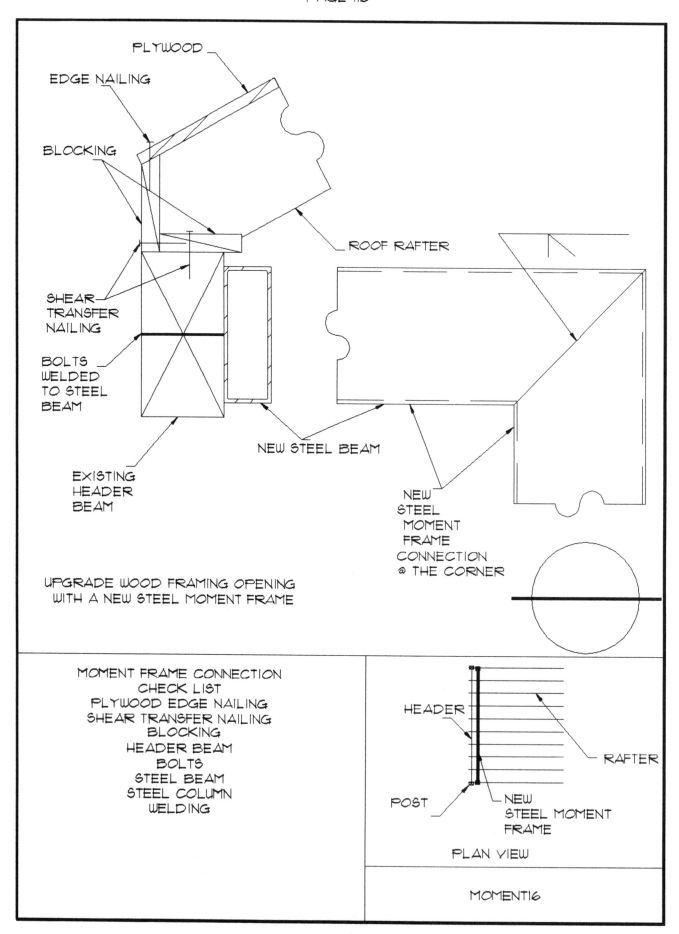

PLYWOOD

EDGE NAILING

BLOCKING

ROOF RAFTER

SHEAR
TRANSFER
NAILING

BOLTS
WELDED
TO STEEL
BEAM

NEW STEEL BEAM

NEW
STEEL
MOMENT
FRAME
CONNECTION
@ THE CORNER

EXISTING
HEADER
BEAM

UPGRADE WOOD FRAMING OPENING
WITH A NEW STEEL MOMENT FRAME

MOMENT FRAME CONNECTION
CHECK LIST
PLYWOOD EDGE NAILING
SHEAR TRANSFER NAILING
BLOCKING
HEADER BEAM
BOLTS
STEEL BEAM
STEEL COLUMN
WELDING

HEADER

RAFTER

POST

NEW
STEEL MOMENT
FRAME

PLAN VIEW

MOMENT16

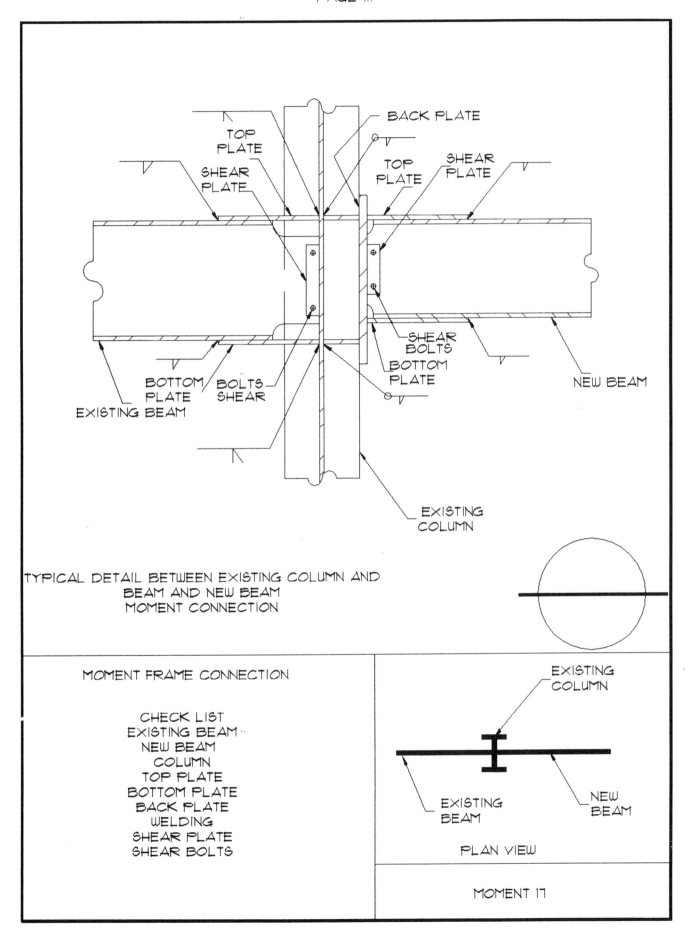

BACK PLATE

TOP PLATE

SHEAR PLATE

TOP PLATE

SHEAR PLATE

SHEAR BOLTS

BOTTOM PLATE

NEW BEAM

BOTTOM PLATE

BOLTS SHEAR

EXISTING BEAM

EXISTING COLUMN

TYPICAL DETAIL BETWEEN EXISTING COLUMN AND
BEAM AND NEW BEAM
MOMENT CONNECTION

MOMENT FRAME CONNECTION

CHECK LIST
EXISTING BEAM
NEW BEAM
COLUMN
TOP PLATE
BOTTOM PLATE
BACK PLATE
WELDING
SHEAR PLATE
SHEAR BOLTS

EXISTING COLUMN

EXISTING BEAM

NEW BEAM

PLAN VIEW

MOMENT 17

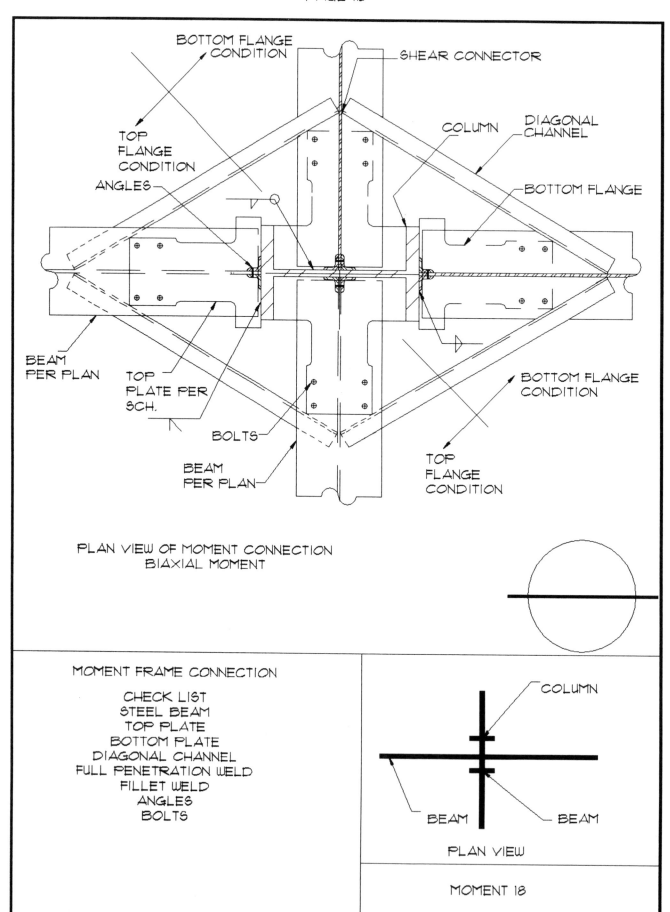

BOTTOM FLANGE
CONDITION

SHEAR CONNECTOR

TOP
FLANGE
CONDITION

COLUMN

DIAGONAL
CHANNEL

ANGLES

BOTTOM FLANGE

BEAM
PER PLAN

TOP
PLATE PER
SCH.

BOLTS

BEAM
PER PLAN

BOTTOM FLANGE
CONDITION

TOP
FLANGE
CONDITION

PLAN VIEW OF MOMENT CONNECTION
BIAXIAL MOMENT

MOMENT FRAME CONNECTION

CHECK LIST
STEEL BEAM
TOP PLATE
BOTTOM PLATE
DIAGONAL CHANNEL
FULL PENETRATION WELD
FILLET WELD
ANGLES
BOLTS

COLUMN

BEAM

BEAM

PLAN VIEW

MOMENT 18

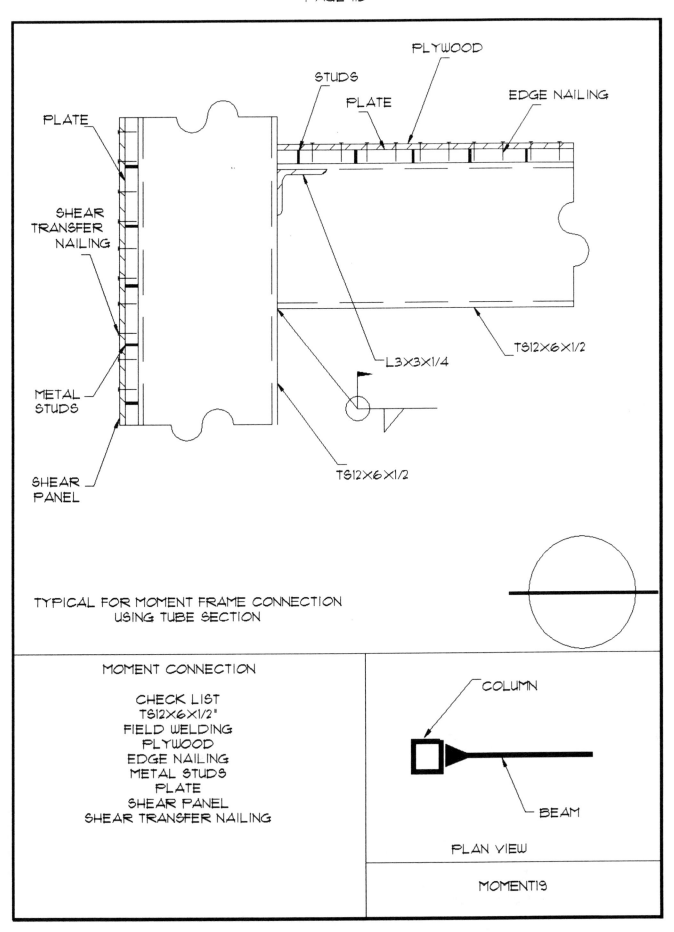

PLATE

STUDS

PLATE

PLYWOOD

EDGE NAILING

PLATE

SHEAR
TRANSFER
NAILING

METAL
STUDS

SHEAR
PANEL

L3X3X1/4

TS12X6X1/2

TS12X6X1/2

TYPICAL FOR MOMENT FRAME CONNECTION
USING TUBE SECTION

MOMENT CONNECTION

CHECK LIST
TS12X6X1/2"
FIELD WELDING
PLYWOOD
EDGE NAILING
METAL STUDS
PLATE
SHEAR PANEL
SHEAR TRANSFER NAILING

COLUMN

BEAM

PLAN VIEW

MOMENT19

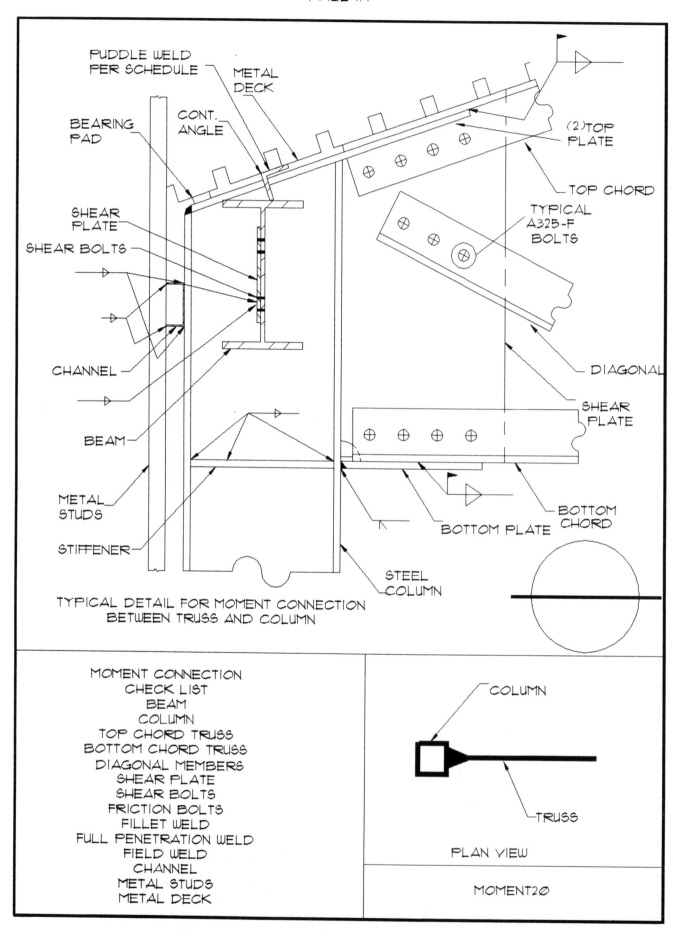

PUDDLE WELD
PER SCHEDULE

METAL
DECK

BEARING
PAD

CONT.
ANGLE

(2) TOP
PLATE

TOP CHORD

SHEAR
PLATE

TYPICAL
A325-F
BOLTS

SHEAR BOLTS

CHANNEL

DIAGONAL

SHEAR
PLATE

BEAM

METAL
STUDS

BOTTOM PLATE

BOTTOM
CHORD

STIFFENER

STEEL
COLUMN

TYPICAL DETAIL FOR MOMENT CONNECTION
BETWEEN TRUSS AND COLUMN

MOMENT CONNECTION
CHECK LIST
BEAM
COLUMN
TOP CHORD TRUSS
BOTTOM CHORD TRUSS
DIAGONAL MEMBERS
SHEAR PLATE
SHEAR BOLTS
FRICTION BOLTS
FILLET WELD
FULL PENETRATION WELD
FIELD WELD
CHANNEL
METAL STUDS
METAL DECK

COLUMN

TRUSS

PLAN VIEW

MOMENT20

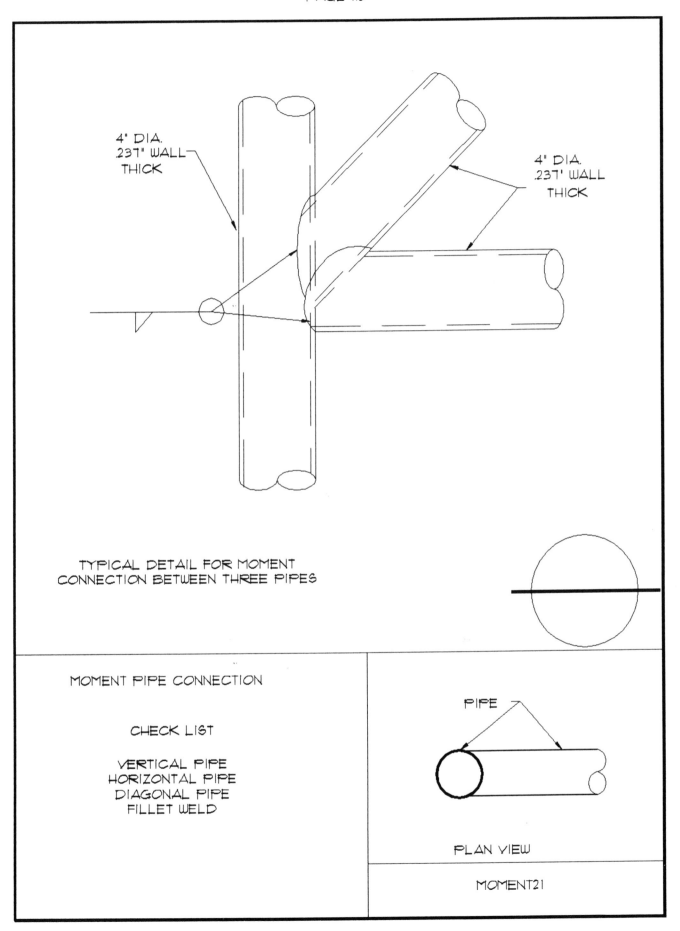

4" DIA.
.237" WALL
THICK

4" DIA.
.237' WALL
THICK

TYPICAL DETAIL FOR MOMENT
CONNECTION BETWEEN THREE PIPES

MOMENT PIPE CONNECTION

CHECK LIST

VERTICAL PIPE
HORIZONTAL PIPE
DIAGONAL PIPE
FILLET WELD

PIPE

PLAN VIEW

MOMENT21

CHAPTER 8

OPENING THROUGH STRUCTURAL MEMBER

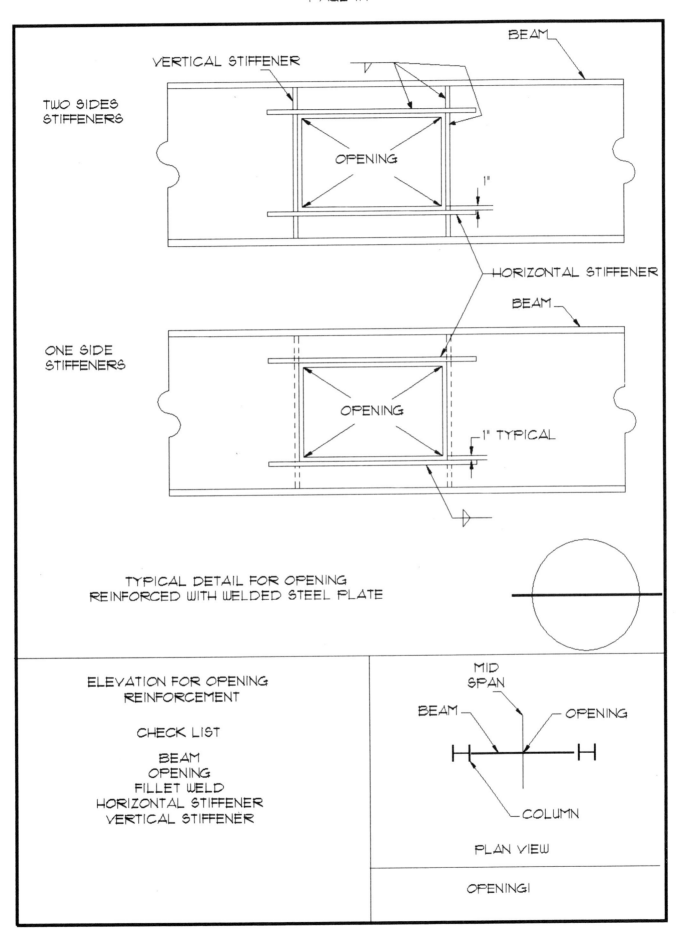

TWO SIDES STIFFENERS

VERTICAL STIFFENER

BEAM

OPENING

1"

HORIZONTAL STIFFENER

BEAM

ONE SIDE STIFFENERS

OPENING

1" TYPICAL

TYPICAL DETAIL FOR OPENING
REINFORCED WITH WELDED STEEL PLATE

ELEVATION FOR OPENING
REINFORCEMENT

CHECK LIST

BEAM
OPENING
FILLET WELD
HORIZONTAL STIFFENER
VERTICAL STIFFENER

MID SPAN

BEAM

OPENING

COLUMN

PLAN VIEW

OPENINGI

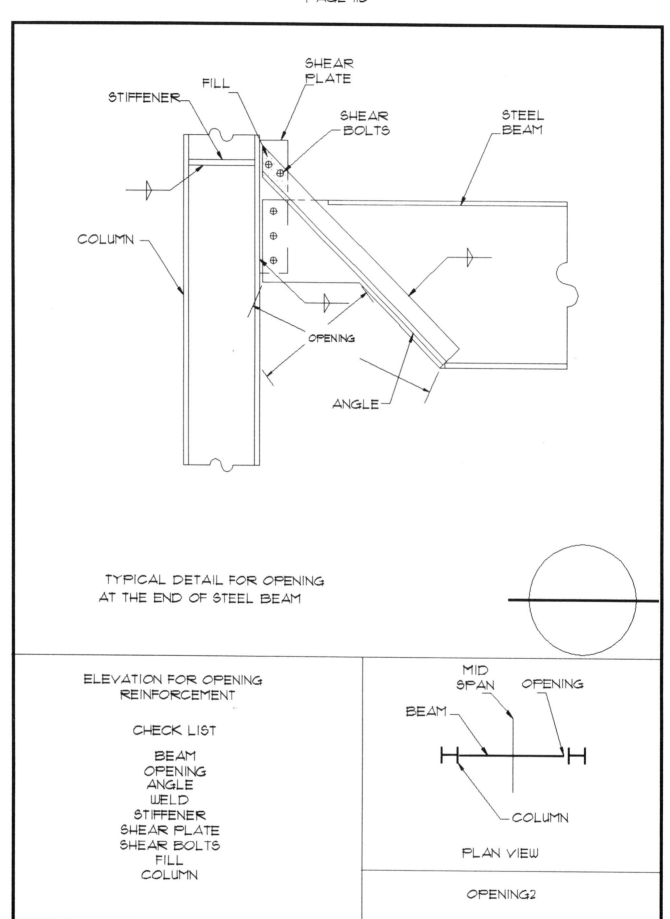

STIFFENER

FILL

SHEAR
PLATE

SHEAR
BOLTS

STEEL
BEAM

COLUMN

OPENING

ANGLE

TYPICAL DETAIL FOR OPENING
AT THE END OF STEEL BEAM

ELEVATION FOR OPENING
REINFORCEMENT

CHECK LIST

BEAM
OPENING
ANGLE
WELD
STIFFENER
SHEAR PLATE
SHEAR BOLTS
FILL
COLUMN

MID
SPAN

OPENING

BEAM

COLUMN

PLAN VIEW

OPENING2

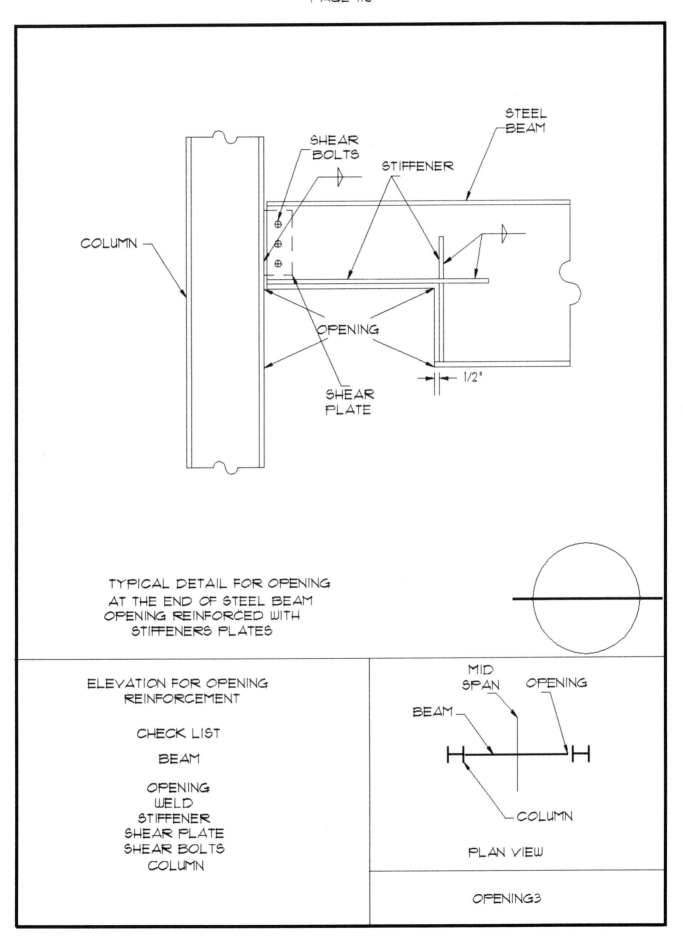

STEEL
BEAM

SHEAR
BOLTS

STIFFENER

COLUMN

OPENING

SHEAR
PLATE

1/2"

TYPICAL DETAIL FOR OPENING
AT THE END OF STEEL BEAM
OPENING REINFORCED WITH
STIFFENERS PLATES

ELEVATION FOR OPENING
REINFORCEMENT

CHECK LIST

BEAM

OPENING
WELD
STIFFENER
SHEAR PLATE
SHEAR BOLTS
COLUMN

MID
SPAN   OPENING

BEAM

COLUMN

PLAN VIEW

OPENING3

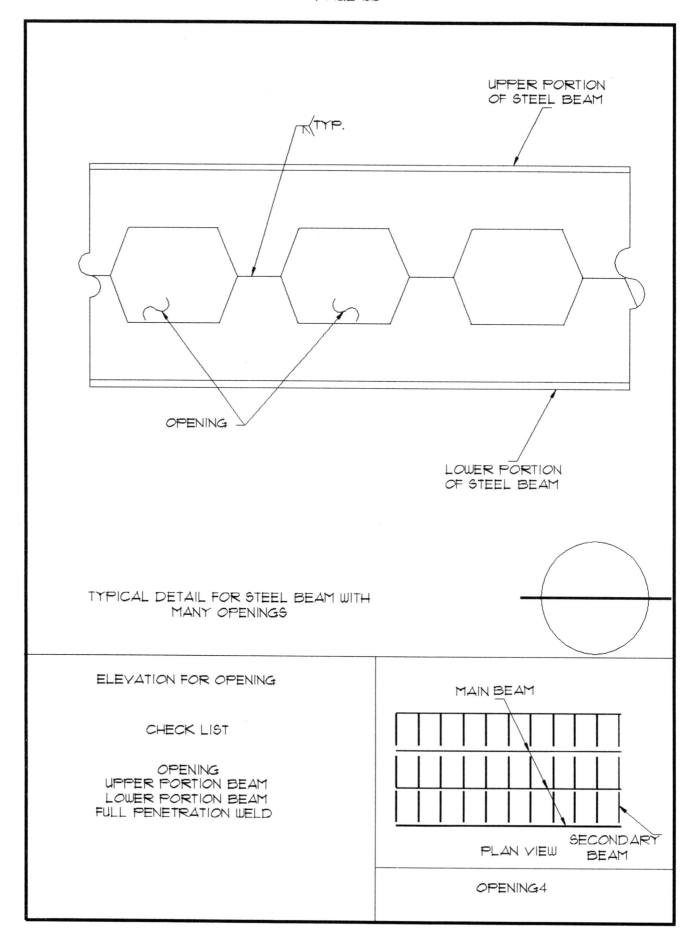

TYP.

UPPER PORTION
OF STEEL BEAM

OPENING

LOWER PORTION
OF STEEL BEAM

TYPICAL DETAIL FOR STEEL BEAM WITH
MANY OPENINGS

ELEVATION FOR OPENING

CHECK LIST

OPENING
UPPER PORTION BEAM
LOWER PORTION BEAM
FULL PENETRATION WELD

MAIN BEAM

PLAN VIEW

SECONDARY
BEAM

OPENING4

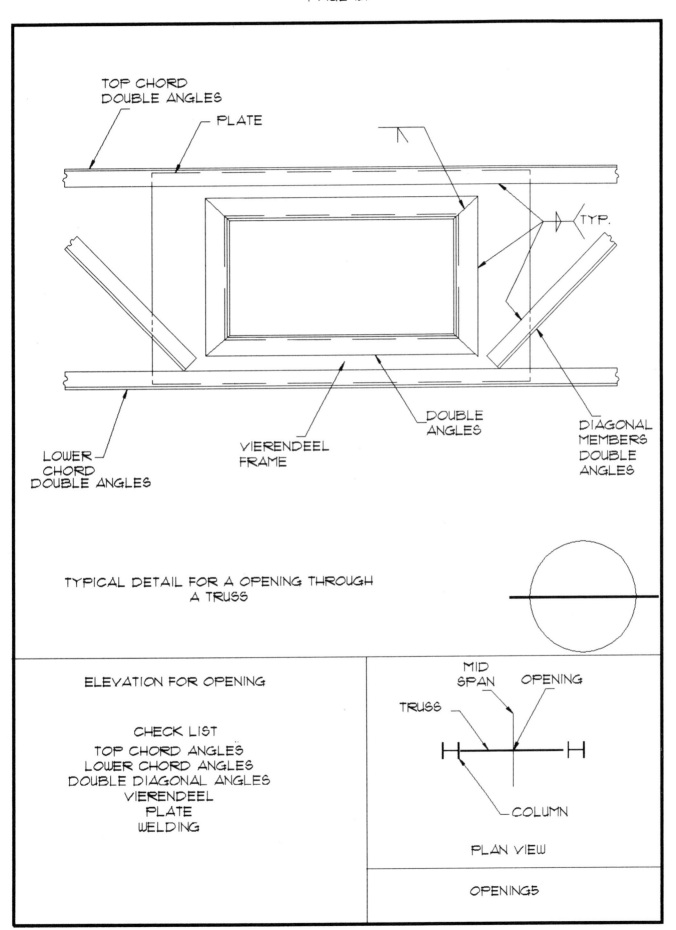

TOP CHORD
DOUBLE ANGLES

PLATE

TYP.

LOWER
CHORD
DOUBLE ANGLES

VIERENDEEL
FRAME

DOUBLE
ANGLES

DIAGONAL
MEMBERS
DOUBLE
ANGLES

TYPICAL DETAIL FOR A OPENING THROUGH
A TRUSS

ELEVATION FOR OPENING

CHECK LIST
TOP CHORD ANGLES
LOWER CHORD ANGLES
DOUBLE DIAGONAL ANGLES
VIERENDEEL
PLATE
WELDING

MID
SPAN   OPENING

TRUSS

COLUMN

PLAN VIEW

OPENINGS

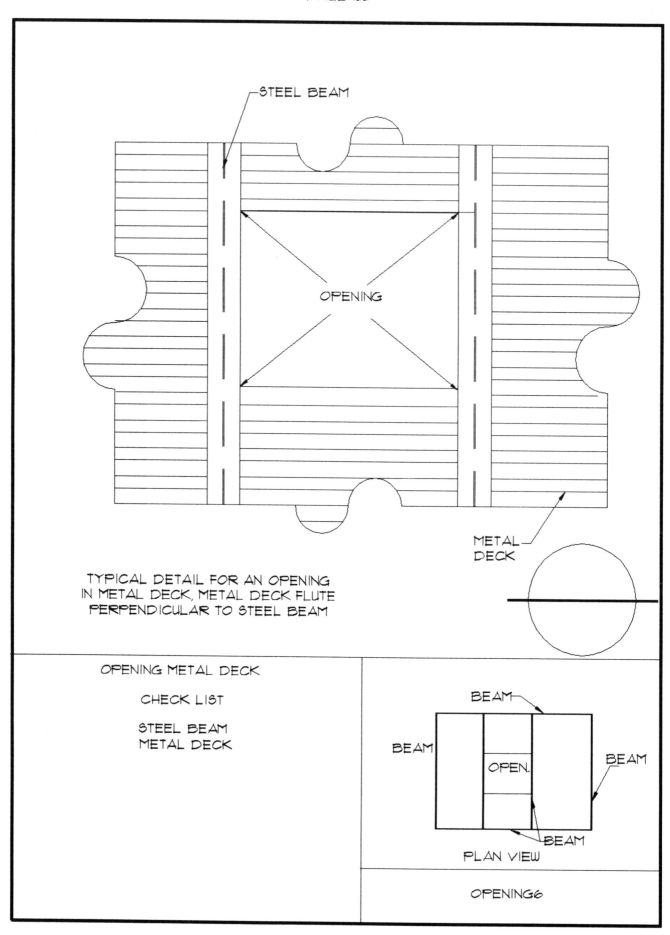

STEEL BEAM

OPENING

METAL DECK

TYPICAL DETAIL FOR AN OPENING
IN METAL DECK, METAL DECK FLUTE
PERPENDICULAR TO STEEL BEAM

OPENING METAL DECK

CHECK LIST

STEEL BEAM
METAL DECK

BEAM

BEAM

OPEN.

BEAM

BEAM

PLAN VIEW

OPENING6

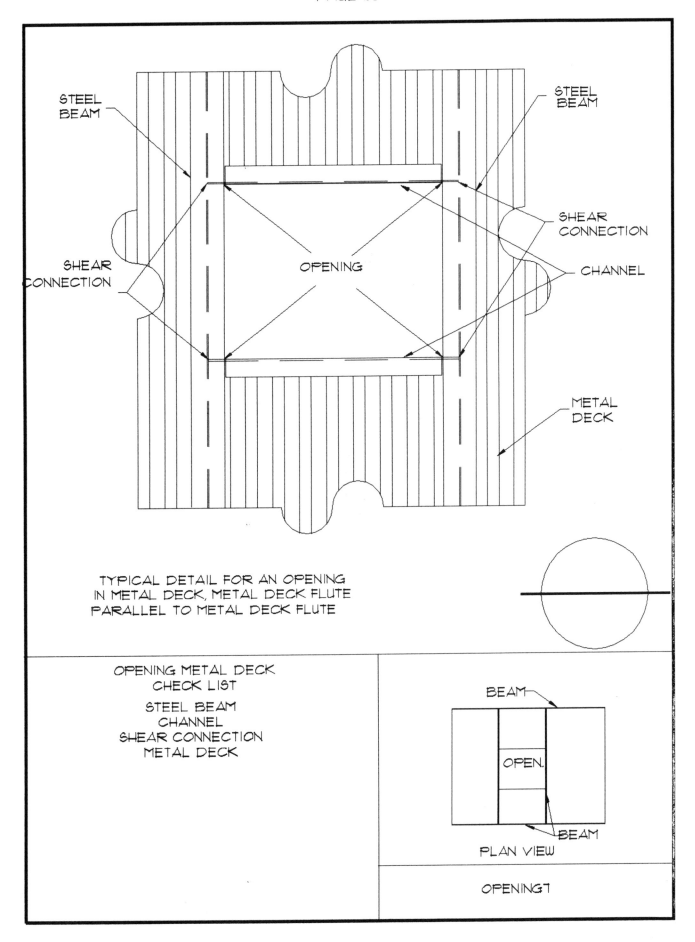

STEEL
BEAM

STEEL
BEAM

SHEAR
CONNECTION

SHEAR
CONNECTION

OPENING

CHANNEL

METAL
DECK

TYPICAL DETAIL FOR AN OPENING
IN METAL DECK, METAL DECK FLUTE
PARALLEL TO METAL DECK FLUTE

OPENING METAL DECK
CHECK LIST

STEEL BEAM
CHANNEL
SHEAR CONNECTION
METAL DECK

BEAM

OPEN.

BEAM

PLAN VIEW

OPENING7

CHAPTER 9

PIN STEEL CONNECTION

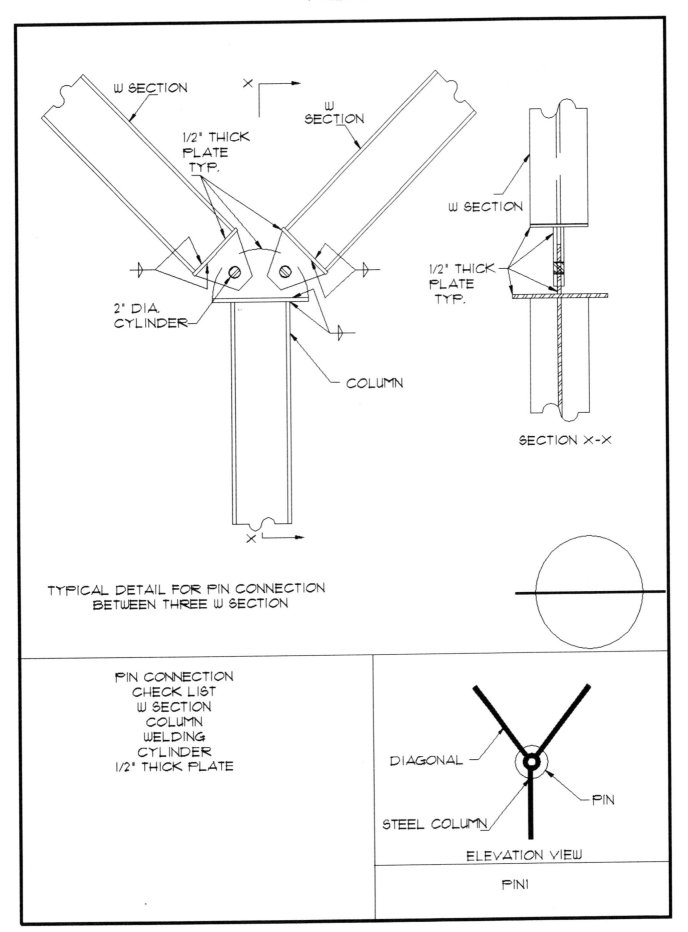

W SECTION

W SECTION

1/2" THICK PLATE TYP.

W SECTION

1/2" THICK PLATE TYP.

2" DIA. CYLINDER

COLUMN

SECTION X-X

TYPICAL DETAIL FOR PIN CONNECTION
BETWEEN THREE W SECTION

PIN CONNECTION
CHECK LIST
W SECTION
COLUMN
WELDING
CYLINDER
1/2" THICK PLATE

DIAGONAL

PIN

STEEL COLUMN

ELEVATION VIEW

PIN1

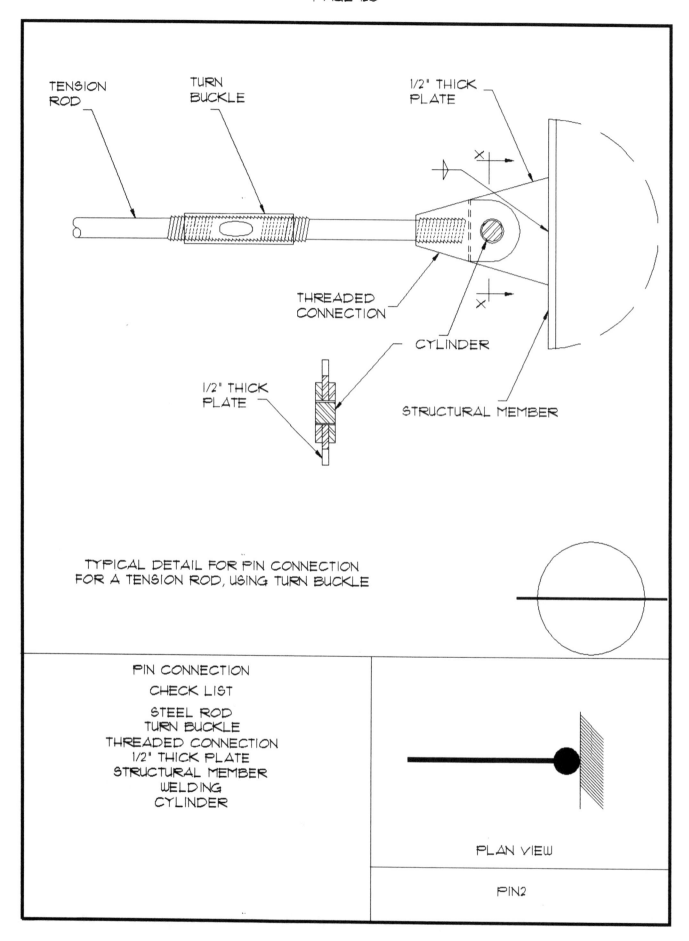

TENSION
ROD

TURN
BUCKLE

1/2" THICK
PLATE

THREADED
CONNECTION

1/2" THICK
PLATE

CYLINDER

STRUCTURAL MEMBER

TYPICAL DETAIL FOR PIN CONNECTION
FOR A TENSION ROD, USING TURN BUCKLE

PIN CONNECTION
CHECK LIST

STEEL ROD
TURN BUCKLE
THREADED CONNECTION
1/2" THICK PLATE
STRUCTURAL MEMBER
WELDING
CYLINDER

PLAN VIEW

PIN2

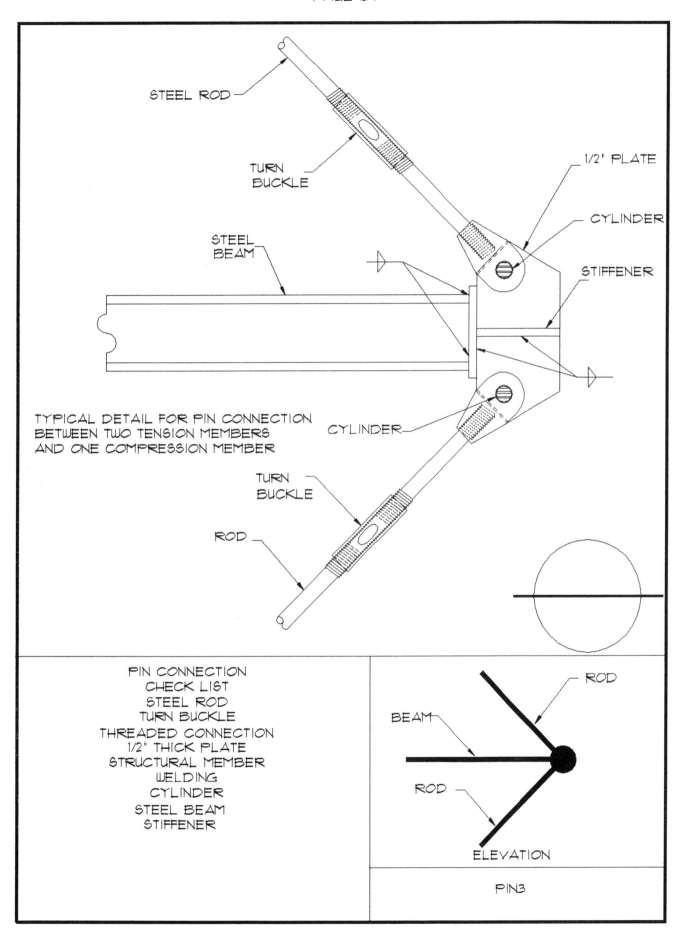

STEEL ROD

TURN
BUCKLE

1/2' PLATE

CYLINDER

STIFFENER

STEEL
BEAM

TYPICAL DETAIL FOR PIN CONNECTION
BETWEEN TWO TENSION MEMBERS
AND ONE COMPRESSION MEMBER

CYLINDER

TURN
BUCKLE

ROD

PIN CONNECTION
CHECK LIST
STEEL ROD
TURN BUCKLE
THREADED CONNECTION
1/2" THICK PLATE
STRUCTURAL MEMBER
WELDING
CYLINDER
STEEL BEAM
STIFFENER

ROD

BEAM

ROD

ELEVATION

PIN3

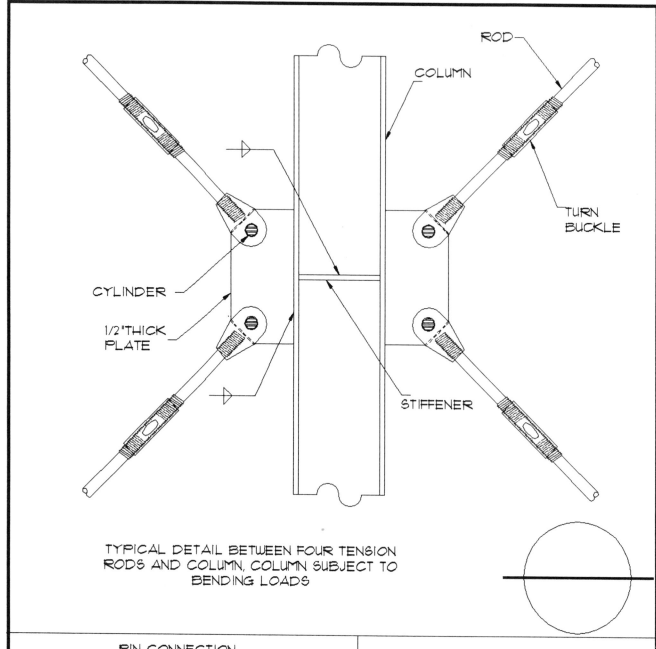

ROD

COLUMN

TURN BUCKLE

CYLINDER

1/2"THICK PLATE

STIFFENER

TYPICAL DETAIL BETWEEN FOUR TENSION RODS AND COLUMN, COLUMN SUBJECT TO BENDING LOADS

PIN CONNECTION
CHECK LIST
STEEL ROD
TURN BUCKLE
THREADED CONNECTION
1/2" THICK PLATE
STRUCTURAL MEMBER
WELDING
CYLINDER
COLUMN
STIFFENER

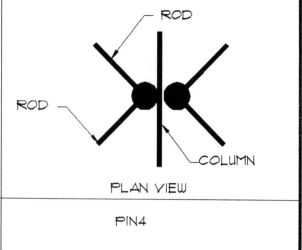

ROD

ROD

COLUMN

PLAN VIEW

PIN4

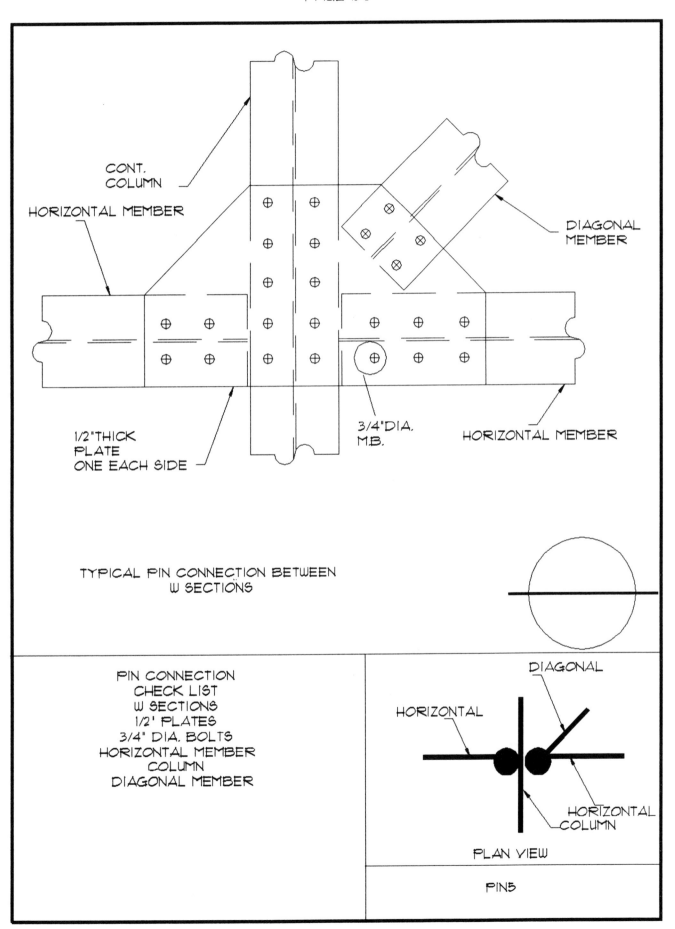

CONT.
COLUMN

HORIZONTAL MEMBER

DIAGONAL
MEMBER

1/2"THICK
PLATE
ONE EACH SIDE

3/4"DIA.
M.B.

HORIZONTAL MEMBER

TYPICAL PIN CONNECTION BETWEEN
W SECTIONS

PIN CONNECTION
CHECK LIST
W SECTIONS
1/2' PLATES
3/4" DIA. BOLTS
HORIZONTAL MEMBER
COLUMN
DIAGONAL MEMBER

DIAGONAL

HORIZONTAL

HORIZONTAL
COLUMN

PLAN VIEW

PIN5

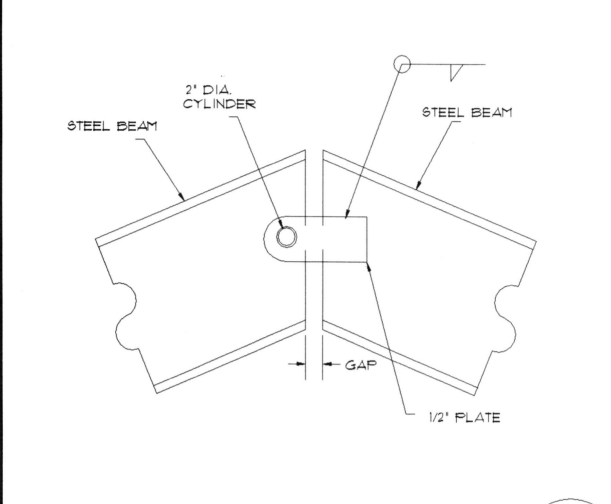

STEEL BEAM

2" DIA. CYLINDER

STEEL BEAM

GAP

1/2" PLATE

TYPICAL DETAIL FOR PIN CONNECTION
BETWEEN TWO SLOPED BEAM

PIN CONNECTION
CHECK LIST

STEEL BEAM
1/2" STEEL PLATE
WELDING
2" DIA. CYLINDER
GAP

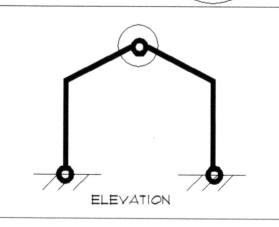

ELEVATION

PIN6

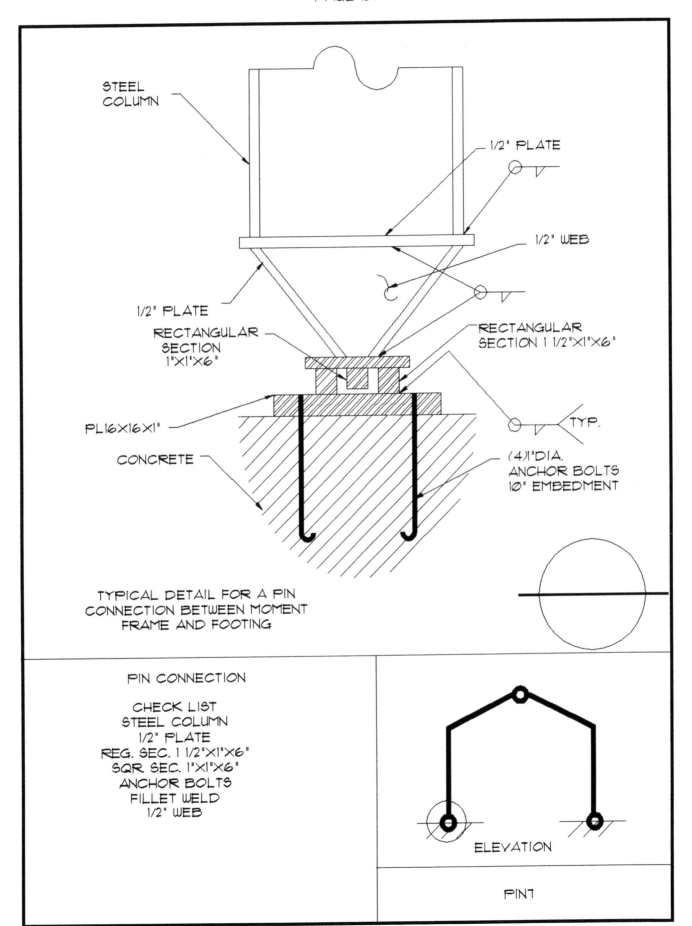

CHAPTER 10

ROOF STRUCTURAL STEEL DETAILS

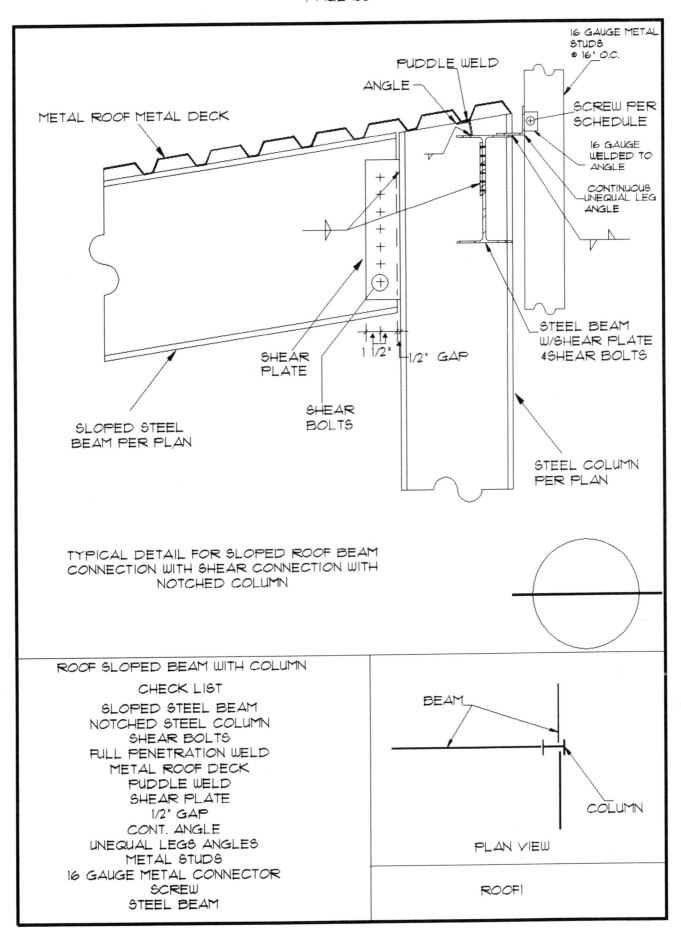

METAL ROOF METAL DECK

ANGLE

PUDDLE WELD

16 GAUGE METAL STUDS @ 16' O.C.

SCREW PER SCHEDULE

16 GAUGE WELDED TO ANGLE

CONTINUOUS UNEQUAL LEG ANGLE

SHEAR PLATE

SHEAR BOLTS

1 1/2"

1/2" GAP

SLOPED STEEL BEAM PER PLAN

STEEL BEAM W/SHEAR PLATE & SHEAR BOLTS

STEEL COLUMN PER PLAN

TYPICAL DETAIL FOR SLOPED ROOF BEAM CONNECTION WITH SHEAR CONNECTION WITH NOTCHED COLUMN

ROOF SLOPED BEAM WITH COLUMN

CHECK LIST

SLOPED STEEL BEAM
NOTCHED STEEL COLUMN
SHEAR BOLTS
FULL PENETRATION WELD
METAL ROOF DECK
PUDDLE WELD
SHEAR PLATE
1/2" GAP
CONT. ANGLE
UNEQUAL LEGS ANGLES
METAL STUDS
16 GAUGE METAL CONNECTOR
SCREW
STEEL BEAM

BEAM

COLUMN

PLAN VIEW

ROOF1

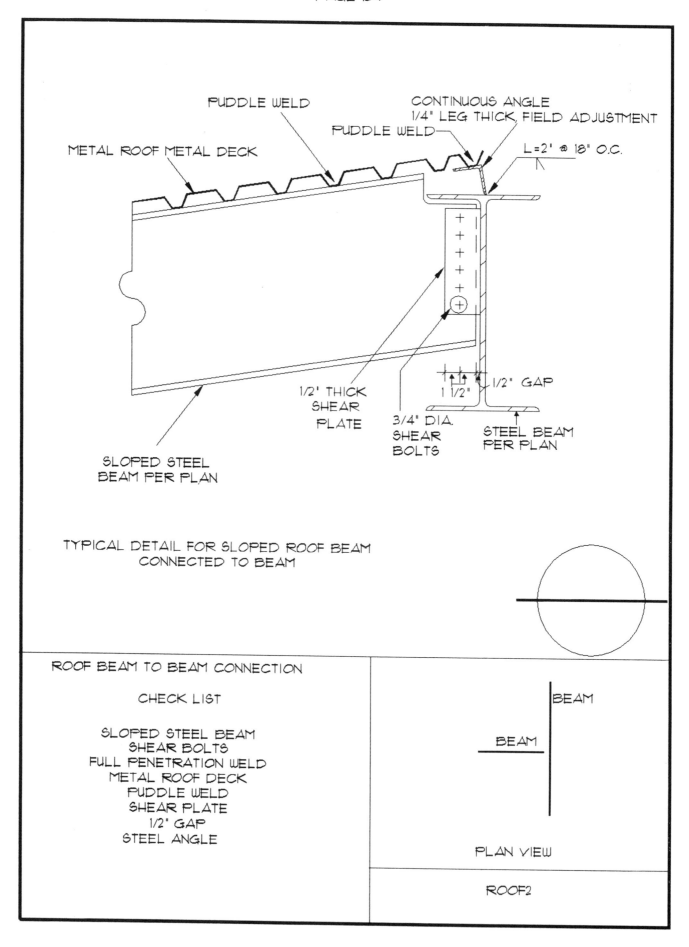

PUDDLE WELD

METAL ROOF METAL DECK

CONTINUOUS ANGLE
1/4" LEG THICK, FIELD ADJUSTMENT

PUDDLE WELD

L=2' @ 18" O.C.

1/2' THICK
SHEAR
PLATE

3/4" DIA.
SHEAR
BOLTS

1 1/2"

1/2" GAP

STEEL BEAM
PER PLAN

SLOPED STEEL
BEAM PER PLAN

TYPICAL DETAIL FOR SLOPED ROOF BEAM
CONNECTED TO BEAM

ROOF BEAM TO BEAM CONNECTION

CHECK LIST

SLOPED STEEL BEAM
SHEAR BOLTS
FULL PENETRATION WELD
METAL ROOF DECK
PUDDLE WELD
SHEAR PLATE
1/2" GAP
STEEL ANGLE

BEAM

BEAM

PLAN VIEW

ROOF2

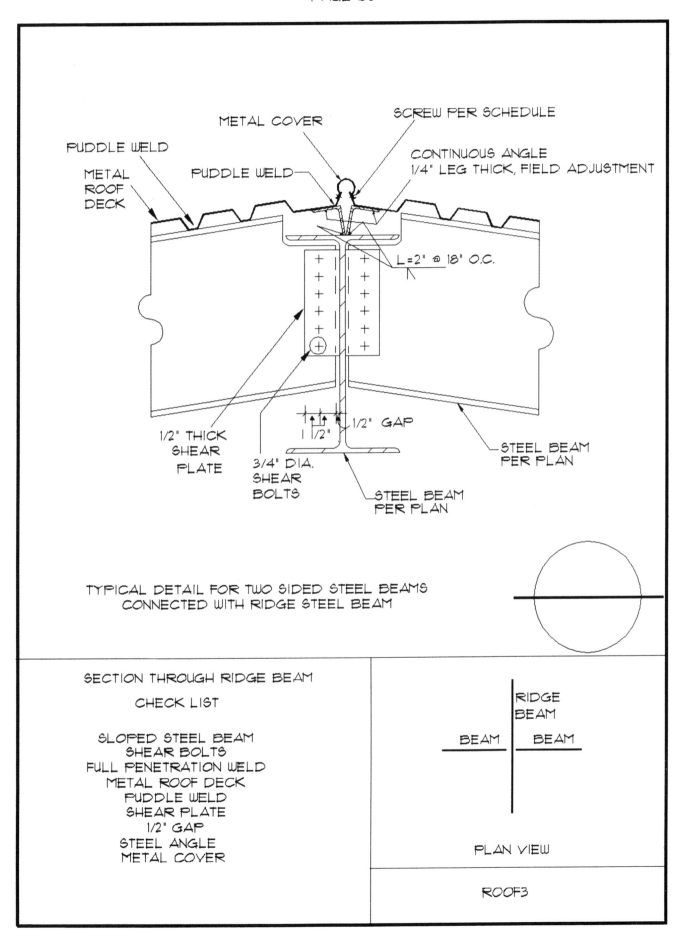

METAL COVER

SCREW PER SCHEDULE

PUDDLE WELD

PUDDLE WELD

CONTINUOUS ANGLE
1/4" LEG THICK, FIELD ADJUSTMENT

METAL
ROOF
DECK

L=2" @ 18" O.C.

1/2" GAP

1 1/2"

1/2" THICK
SHEAR
PLATE

3/4" DIA.
SHEAR
BOLTS

STEEL BEAM
PER PLAN

STEEL BEAM
PER PLAN

TYPICAL DETAIL FOR TWO SIDED STEEL BEAMS
CONNECTED WITH RIDGE STEEL BEAM

SECTION THROUGH RIDGE BEAM

CHECK LIST

SLOPED STEEL BEAM
SHEAR BOLTS
FULL PENETRATION WELD
METAL ROOF DECK
PUDDLE WELD
SHEAR PLATE
1/2" GAP
STEEL ANGLE
METAL COVER

RIDGE
BEAM

BEAM   BEAM

PLAN VIEW

ROOF3

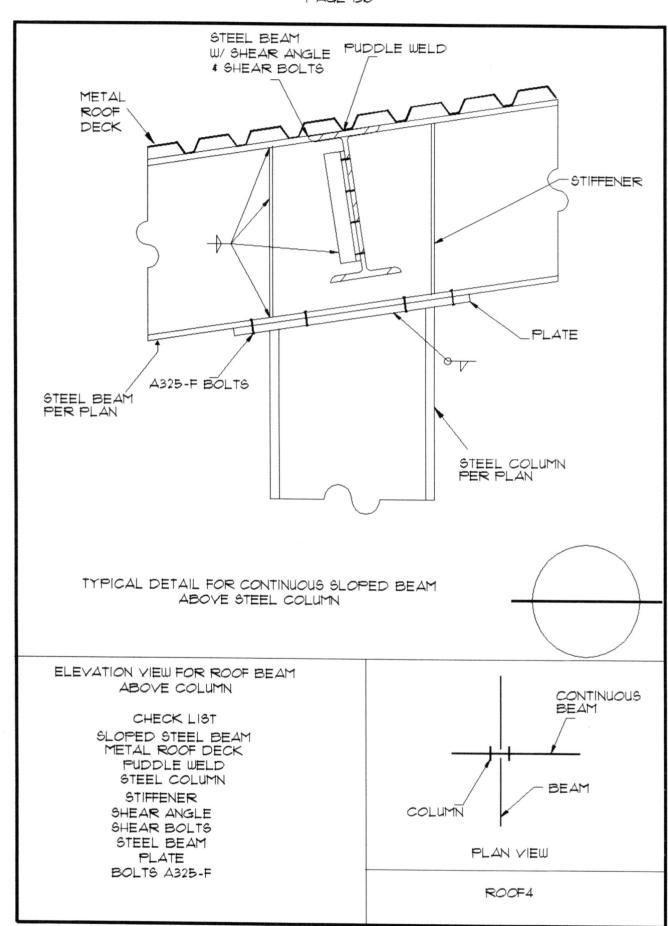

STEEL BEAM
W/ SHEAR ANGLE
& SHEAR BOLTS

PUDDLE WELD

METAL
ROOF
DECK

STIFFENER

STEEL BEAM
PER PLAN

A325-F BOLTS

PLATE

STEEL COLUMN
PER PLAN

TYPICAL DETAIL FOR CONTINUOUS SLOPED BEAM
ABOVE STEEL COLUMN

ELEVATION VIEW FOR ROOF BEAM
ABOVE COLUMN

CHECK LIST
SLOPED STEEL BEAM
METAL ROOF DECK
PUDDLE WELD
STEEL COLUMN
STIFFENER
SHEAR ANGLE
SHEAR BOLTS
STEEL BEAM
PLATE
BOLTS A325-F

CONTINUOUS
BEAM

BEAM

COLUMN

PLAN VIEW

ROOF4

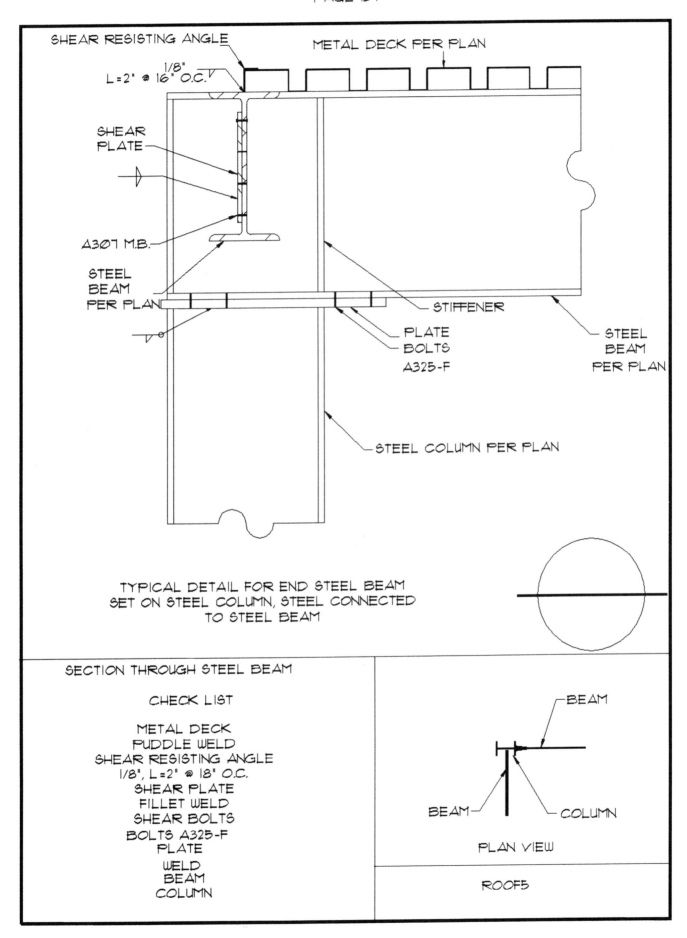

SHEAR RESISTING ANGLE

METAL DECK PER PLAN

L=2" @ 16" O.C. 1/8"

SHEAR PLATE

A307 M.B.

STEEL BEAM PER PLAN

STIFFENER

PLATE
BOLTS
A325-F

STEEL BEAM PER PLAN

STEEL COLUMN PER PLAN

TYPICAL DETAIL FOR END STEEL BEAM
SET ON STEEL COLUMN, STEEL CONNECTED
TO STEEL BEAM

SECTION THROUGH STEEL BEAM

CHECK LIST

METAL DECK
PUDDLE WELD
SHEAR RESISTING ANGLE
1/8", L=2" @ 18" O.C.
SHEAR PLATE
FILLET WELD
SHEAR BOLTS
BOLTS A325-F
PLATE
WELD
BEAM
COLUMN

BEAM

BEAM

COLUMN

PLAN VIEW

ROOF5

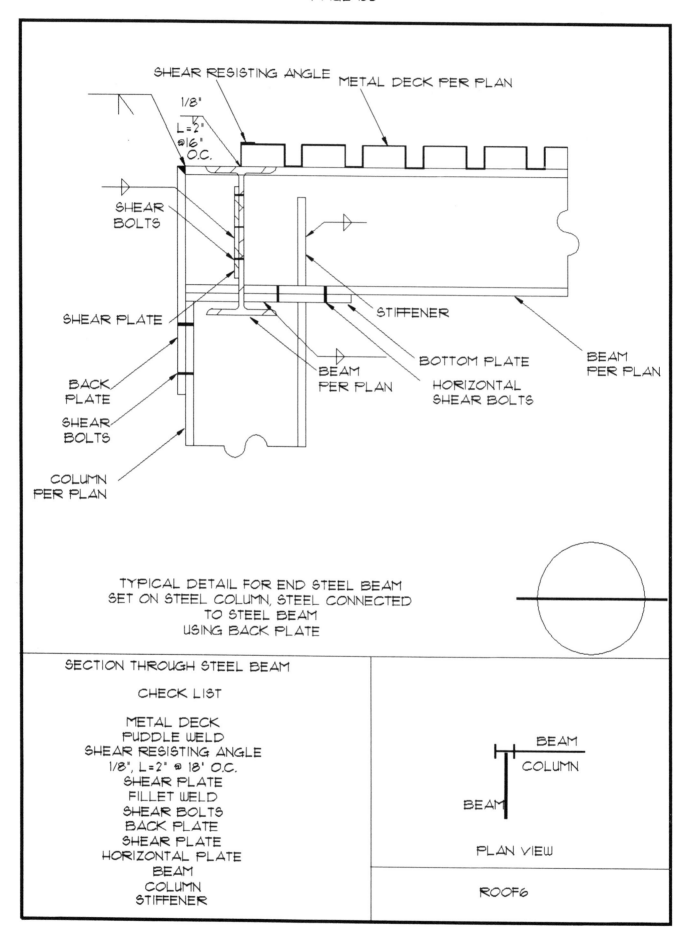

SHEAR RESISTING ANGLE METAL DECK PER PLAN

1/8"
L=2"
@16"
O.C.

SHEAR
BOLTS

SHEAR PLATE

STIFFENER

BACK
PLATE

BEAM
PER PLAN

BOTTOM PLATE

HORIZONTAL
SHEAR BOLTS

BEAM
PER PLAN

SHEAR
BOLTS

COLUMN
PER PLAN

TYPICAL DETAIL FOR END STEEL BEAM
SET ON STEEL COLUMN, STEEL CONNECTED
TO STEEL BEAM
USING BACK PLATE

SECTION THROUGH STEEL BEAM

CHECK LIST

METAL DECK
PUDDLE WELD
SHEAR RESISTING ANGLE
1/8", L=2" @ 18' O.C.
SHEAR PLATE
FILLET WELD
SHEAR BOLTS
BACK PLATE
SHEAR PLATE
HORIZONTAL PLATE
BEAM
COLUMN
STIFFENER

BEAM
COLUMN

BEAM

PLAN VIEW

ROOF6

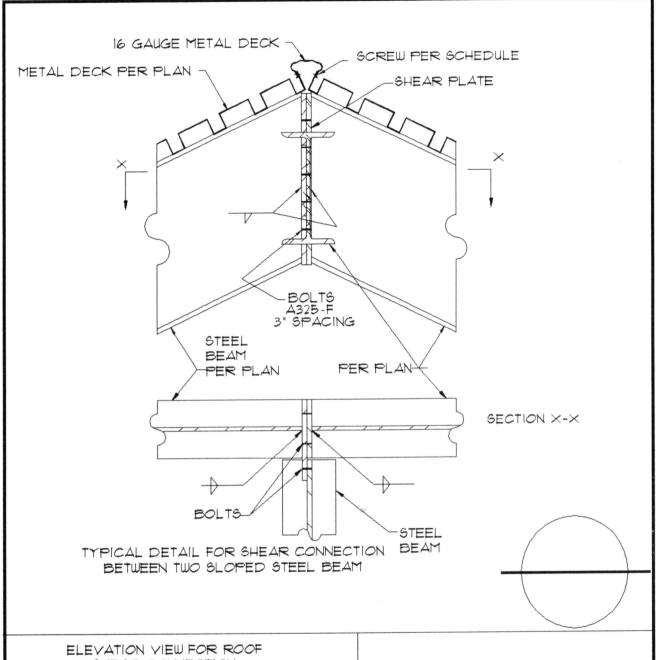

16 GAUGE METAL DECK

SCREW PER SCHEDULE

METAL DECK PER PLAN

SHEAR PLATE

BOLTS
A325-F
3" SPACING

STEEL
BEAM
PER PLAN

PER PLAN

SECTION X-X

BOLTS

STEEL
BEAM

TYPICAL DETAIL FOR SHEAR CONNECTION
BETWEEN TWO SLOPED STEEL BEAM

ELEVATION VIEW FOR ROOF
SHEAR CONNECTION

CHECK LIST

METAL DECK
PUDDLE WELD
FILLET WELD
SHEAR BOLTS
SHEAR PLATES
BEAM

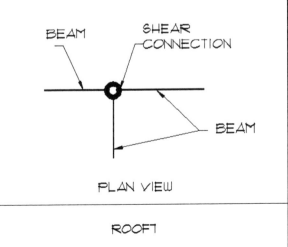

BEAM

SHEAR
CONNECTION

BEAM

PLAN VIEW

ROOF

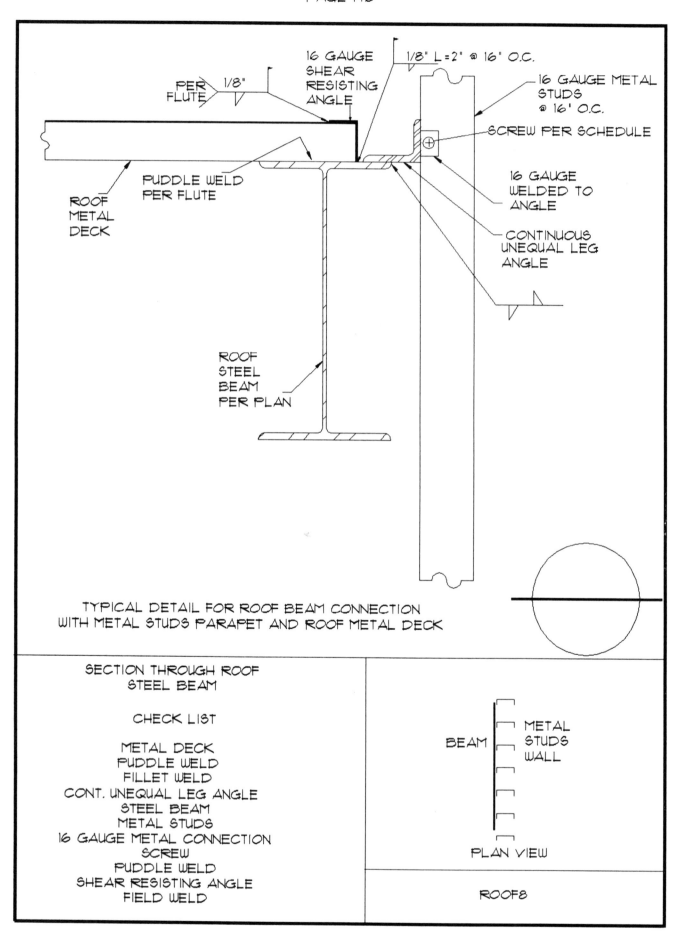

PER FLUTE 1/8"

16 GAUGE SHEAR RESISTING ANGLE

1/8" L=2" @ 16" O.C.

16 GAUGE METAL STUDS @ 16' O.C.

SCREW PER SCHEDULE

16 GAUGE WELDED TO ANGLE

CONTINUOUS UNEQUAL LEG ANGLE

ROOF METAL DECK

PUDDLE WELD PER FLUTE

ROOF STEEL BEAM PER PLAN

TYPICAL DETAIL FOR ROOF BEAM CONNECTION
WITH METAL STUDS PARAPET AND ROOF METAL DECK

SECTION THROUGH ROOF
STEEL BEAM

CHECK LIST

METAL DECK
PUDDLE WELD
FILLET WELD
CONT. UNEQUAL LEG ANGLE
STEEL BEAM
METAL STUDS
16 GAUGE METAL CONNECTION
SCREW
PUDDLE WELD
SHEAR RESISTING ANGLE
FIELD WELD

BEAM

METAL STUDS WALL

PLAN VIEW

ROOF8

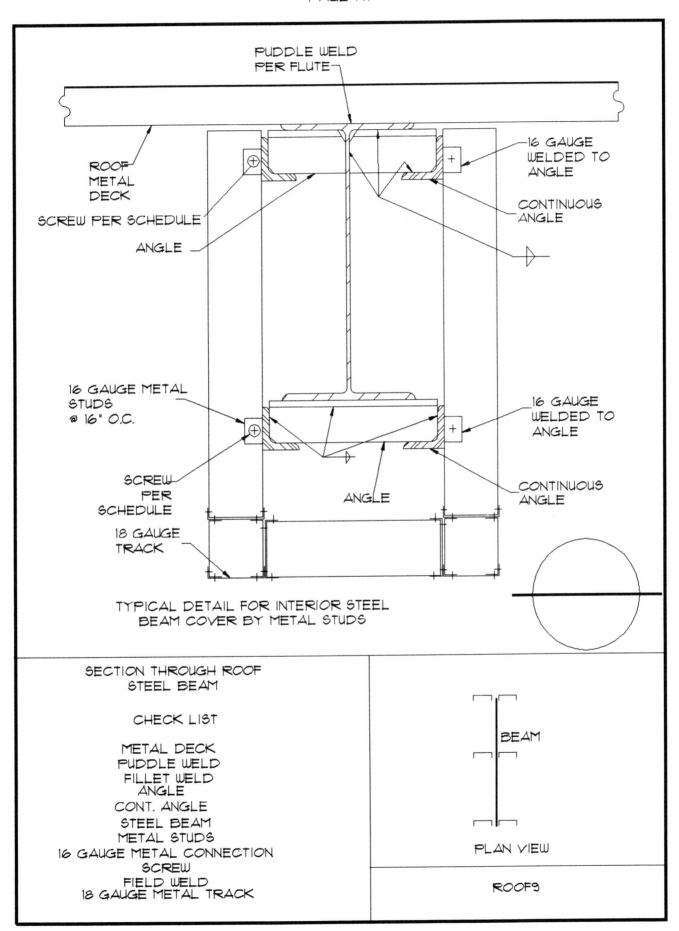

PUDDLE WELD
PER FLUTE

ROOF
METAL
DECK

SCREW PER SCHEDULE

ANGLE

16 GAUGE
WELDED TO
ANGLE

CONTINUOUS
ANGLE

16 GAUGE METAL
STUDS
@ 16" O.C.

16 GAUGE
WELDED TO
ANGLE

SCREW
PER
SCHEDULE

CONTINUOUS
ANGLE

ANGLE

18 GAUGE
TRACK

TYPICAL DETAIL FOR INTERIOR STEEL
BEAM COVER BY METAL STUDS

SECTION THROUGH ROOF
STEEL BEAM

CHECK LIST

METAL DECK
PUDDLE WELD
FILLET WELD
ANGLE
CONT. ANGLE
STEEL BEAM
METAL STUDS
16 GAUGE METAL CONNECTION
SCREW
FIELD WELD
18 GAUGE METAL TRACK

BEAM

PLAN VIEW

ROOF9

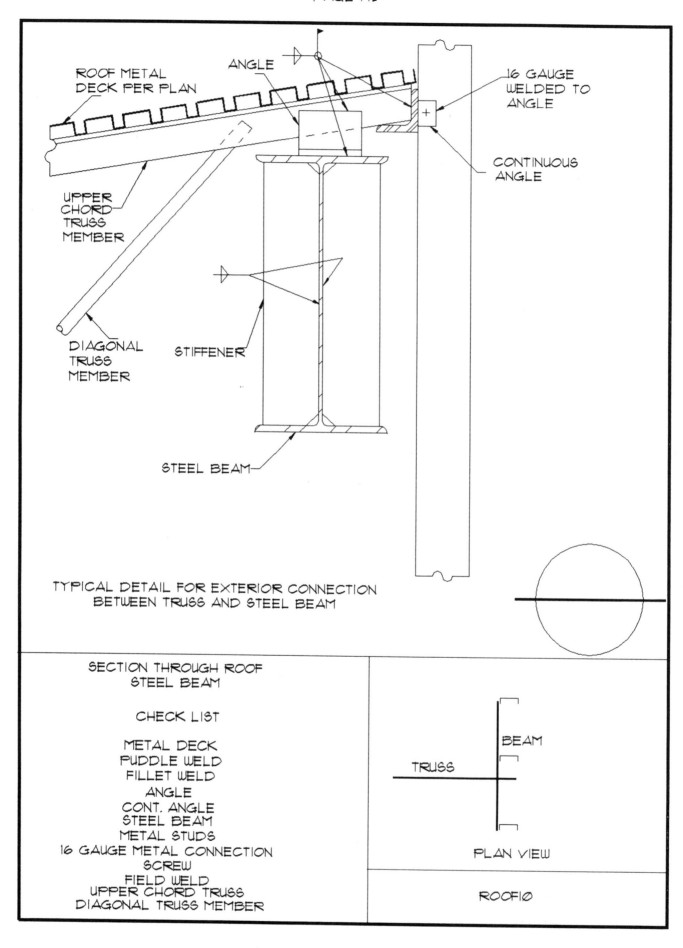

ROOF METAL DECK PER PLAN

ANGLE

16 GAUGE WELDED TO ANGLE

CONTINUOUS ANGLE

UPPER CHORD TRUSS MEMBER

DIAGONAL TRUSS MEMBER

STIFFENER

STEEL BEAM

TYPICAL DETAIL FOR EXTERIOR CONNECTION BETWEEN TRUSS AND STEEL BEAM

SECTION THROUGH ROOF STEEL BEAM

CHECK LIST

METAL DECK
PUDDLE WELD
FILLET WELD
ANGLE
CONT. ANGLE
STEEL BEAM
METAL STUDS
16 GAUGE METAL CONNECTION
SCREW
FIELD WELD
UPPER CHORD TRUSS
DIAGONAL TRUSS MEMBER

BEAM

TRUSS

PLAN VIEW

ROOF10

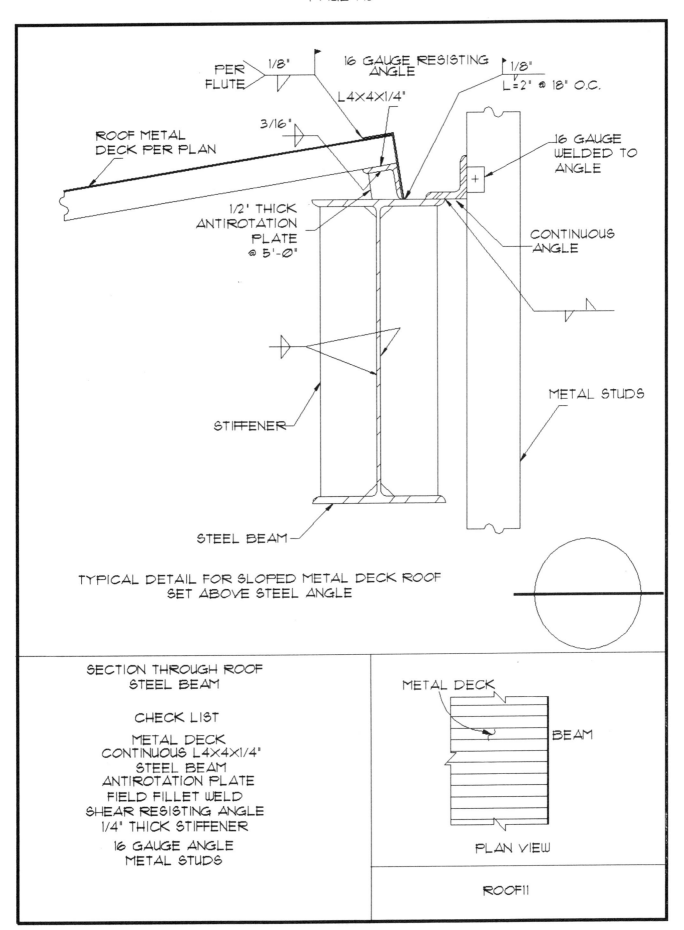

PER FLUTE 1/8"

16 GAUGE RESISTING ANGLE

1/8"
L = 2" @ 18" O.C.

L4X4X1/4"

ROOF METAL DECK PER PLAN

3/16"

16 GAUGE WELDED TO ANGLE

1/2' THICK ANTIROTATION PLATE @ 5'-0"

CONTINUOUS ANGLE

METAL STUDS

STIFFENER

STEEL BEAM

TYPICAL DETAIL FOR SLOPED METAL DECK ROOF
SET ABOVE STEEL ANGLE

SECTION THROUGH ROOF
STEEL BEAM

CHECK LIST

METAL DECK
CONTINUOUS L4X4X1/4"
STEEL BEAM
ANTIROTATION PLATE
FIELD FILLET WELD
SHEAR RESISTING ANGLE
1/4" THICK STIFFENER

16 GAUGE ANGLE
METAL STUDS

METAL DECK

BEAM

PLAN VIEW

ROOF11

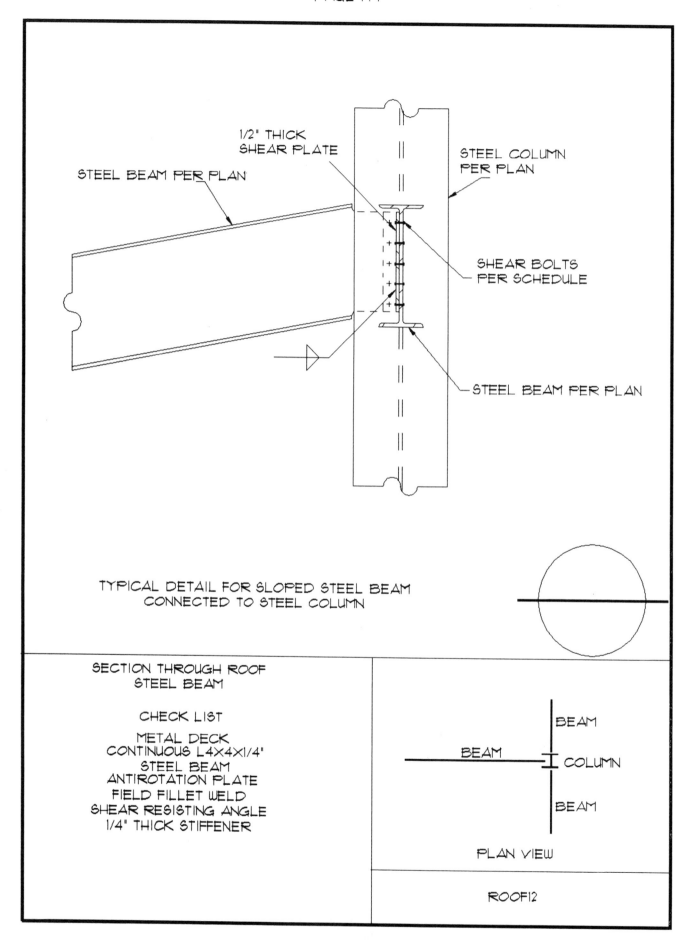

1/2" THICK
SHEAR PLATE

STEEL BEAM PER PLAN

STEEL COLUMN
PER PLAN

SHEAR BOLTS
PER SCHEDULE

STEEL BEAM PER PLAN

TYPICAL DETAIL FOR SLOPED STEEL BEAM
CONNECTED TO STEEL COLUMN

SECTION THROUGH ROOF
STEEL BEAM

CHECK LIST

METAL DECK
CONTINUOUS L4X4X1/4"
STEEL BEAM
ANTIROTATION PLATE
FIELD FILLET WELD
SHEAR RESISTING ANGLE
1/4" THICK STIFFENER

BEAM

BEAM   COLUMN

BEAM

PLAN VIEW

ROOF12

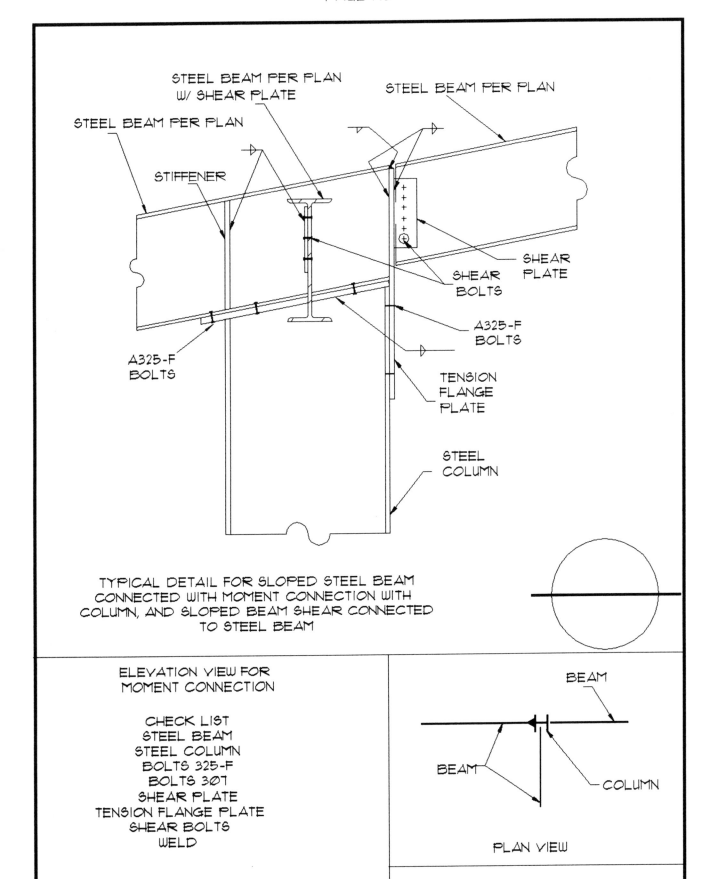

STEEL BEAM PER PLAN
W/ SHEAR PLATE

STEEL BEAM PER PLAN

STEEL BEAM PER PLAN

STIFFENER

SHEAR PLATE

SHEAR BOLTS

A325-F BOLTS

A325-F BOLTS

TENSION FLANGE PLATE

STEEL COLUMN

TYPICAL DETAIL FOR SLOPED STEEL BEAM
CONNECTED WITH MOMENT CONNECTION WITH
COLUMN, AND SLOPED BEAM SHEAR CONNECTED
TO STEEL BEAM

ELEVATION VIEW FOR
MOMENT CONNECTION

CHECK LIST
STEEL BEAM
STEEL COLUMN
BOLTS 325-F
BOLTS 307
SHEAR PLATE
TENSION FLANGE PLATE
SHEAR BOLTS
WELD

BEAM

BEAM

COLUMN

PLAN VIEW

ROOF13

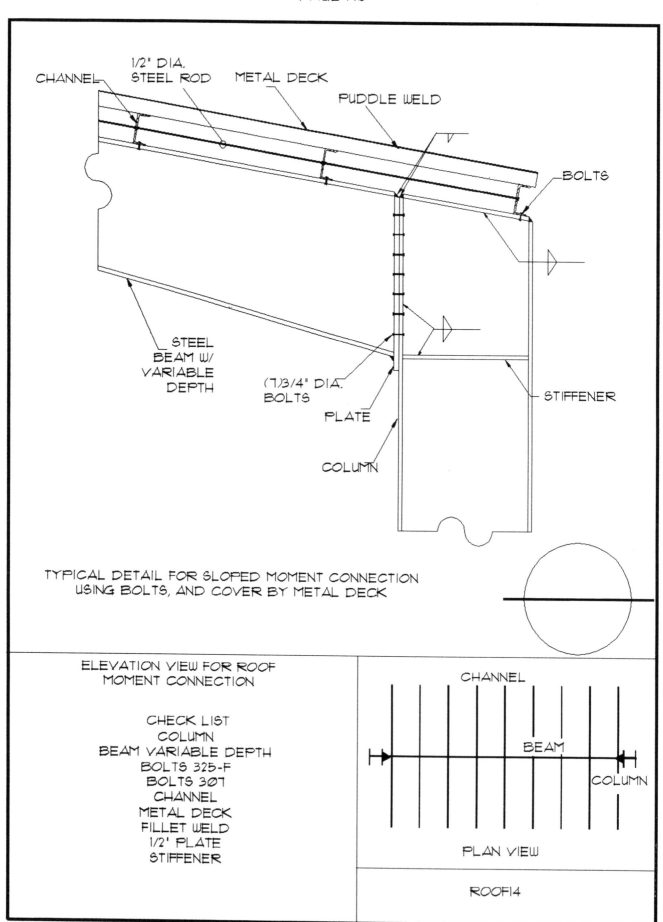

CHANNEL

1/2" DIA. STEEL ROD

METAL DECK

PUDDLE WELD

BOLTS

STEEL BEAM W/ VARIABLE DEPTH

(7)3/4" DIA. BOLTS

PLATE

COLUMN

STIFFENER

TYPICAL DETAIL FOR SLOPED MOMENT CONNECTION
USING BOLTS, AND COVER BY METAL DECK

ELEVATION VIEW FOR ROOF
MOMENT CONNECTION

CHECK LIST
COLUMN
BEAM VARIABLE DEPTH
BOLTS 325-F
BOLTS 307
CHANNEL
METAL DECK
FILLET WELD
1/2' PLATE
STIFFENER

CHANNEL

BEAM

COLUMN

PLAN VIEW

ROOF14

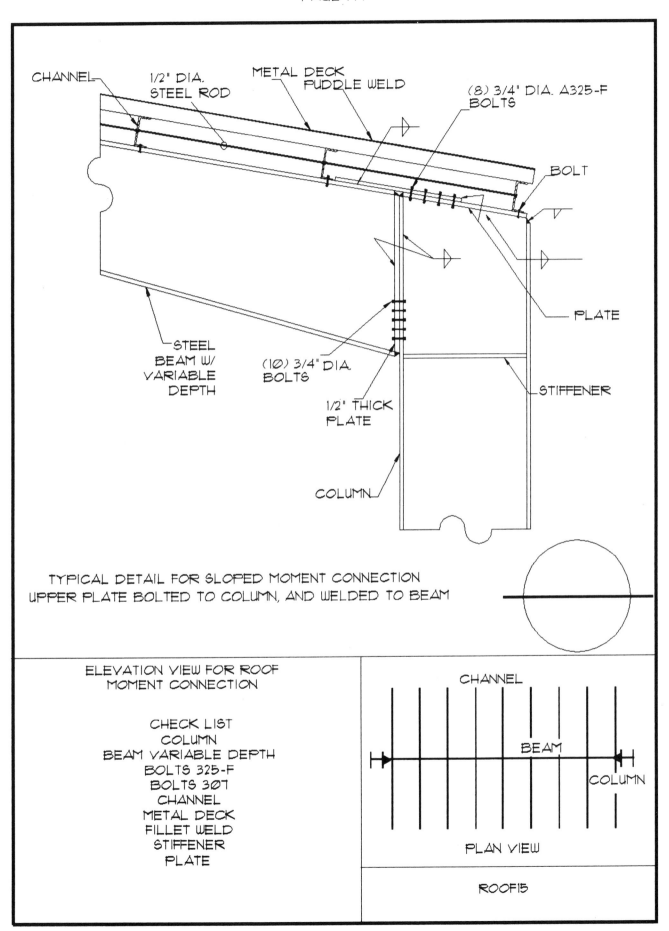

CHANNEL

1/2" DIA. STEEL ROD

METAL DECK PUDDLE WELD

(8) 3/4" DIA. A325-F BOLTS

BOLT

PLATE

STEEL BEAM W/ VARIABLE DEPTH

(10) 3/4" DIA. BOLTS

1/2" THICK PLATE

STIFFENER

COLUMN

TYPICAL DETAIL FOR SLOPED MOMENT CONNECTION
UPPER PLATE BOLTED TO COLUMN, AND WELDED TO BEAM

ELEVATION VIEW FOR ROOF
MOMENT CONNECTION

CHECK LIST
COLUMN
BEAM VARIABLE DEPTH
BOLTS 325-F
BOLTS 307
CHANNEL
METAL DECK
FILLET WELD
STIFFENER
PLATE

CHANNEL

BEAM

COLUMN

PLAN VIEW

ROOF15

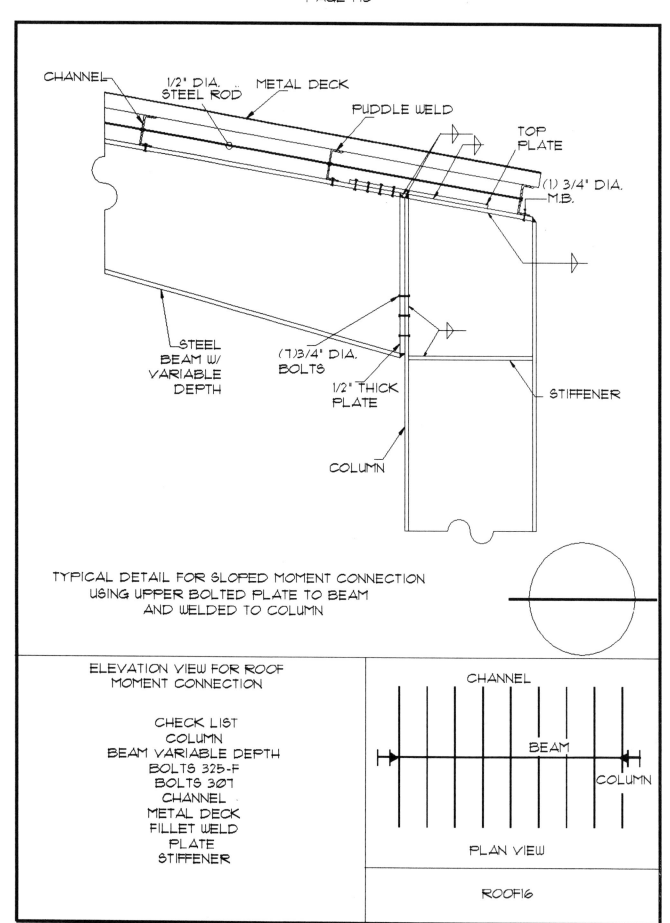

CHANNEL

1/2" DIA. STEEL ROD

METAL DECK

PUDDLE WELD

TOP PLATE

(1) 3/4" DIA. M.B.

STEEL BEAM W/ VARIABLE DEPTH

(7)3/4" DIA. BOLTS

1/2" THICK PLATE

STIFFENER

COLUMN

TYPICAL DETAIL FOR SLOPED MOMENT CONNECTION
USING UPPER BOLTED PLATE TO BEAM
AND WELDED TO COLUMN

ELEVATION VIEW FOR ROOF
MOMENT CONNECTION

CHECK LIST
COLUMN
BEAM VARIABLE DEPTH
BOLTS 325-F
BOLTS 307
CHANNEL
METAL DECK
FILLET WELD
PLATE
STIFFENER

CHANNEL

BEAM

COLUMN

PLAN VIEW

ROOF16

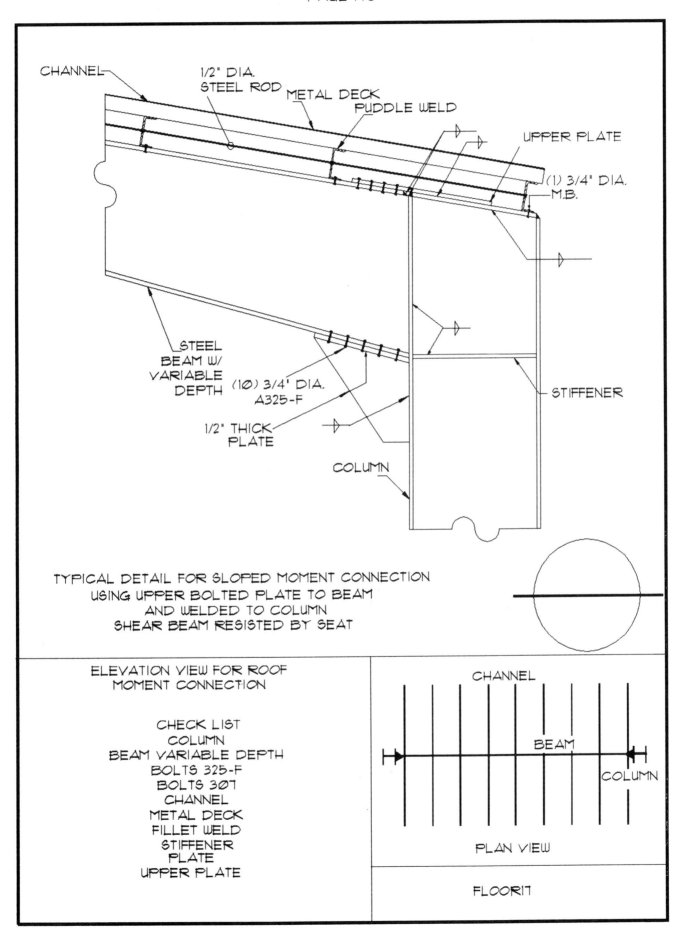

CHANNEL

1/2" DIA. STEEL ROD

METAL DECK

PUDDLE WELD

UPPER PLATE

(1) 3/4" DIA. M.B.

STEEL BEAM W/ VARIABLE DEPTH

(10) 3/4' DIA. A325-F

1/2" THICK PLATE

COLUMN

STIFFENER

TYPICAL DETAIL FOR SLOPED MOMENT CONNECTION
USING UPPER BOLTED PLATE TO BEAM
AND WELDED TO COLUMN
SHEAR BEAM RESISTED BY SEAT

ELEVATION VIEW FOR ROOF
MOMENT CONNECTION

CHECK LIST
COLUMN
BEAM VARIABLE DEPTH
BOLTS 325-F
BOLTS 307
CHANNEL
METAL DECK
FILLET WELD
STIFFENER
PLATE
UPPER PLATE

CHANNEL

BEAM

COLUMN

PLAN VIEW

FLOOR17

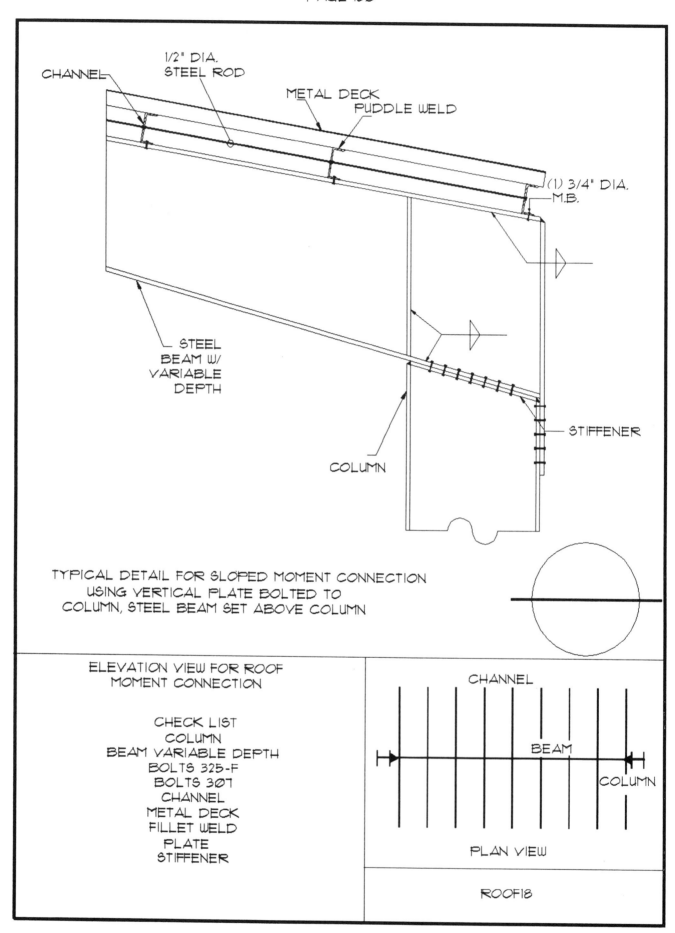

CHANNEL

1/2" DIA.
STEEL ROD

METAL DECK

PUDDLE WELD

(1) 3/4" DIA.
M.B.

STEEL
BEAM W/
VARIABLE
DEPTH

COLUMN

STIFFENER

TYPICAL DETAIL FOR SLOPED MOMENT CONNECTION
USING VERTICAL PLATE BOLTED TO
COLUMN, STEEL BEAM SET ABOVE COLUMN

ELEVATION VIEW FOR ROOF
MOMENT CONNECTION

CHECK LIST
COLUMN
BEAM VARIABLE DEPTH
BOLTS 325-F
BOLTS 307
CHANNEL
METAL DECK
FILLET WELD
PLATE
STIFFENER

CHANNEL

BEAM

COLUMN

PLAN VIEW

ROOF18

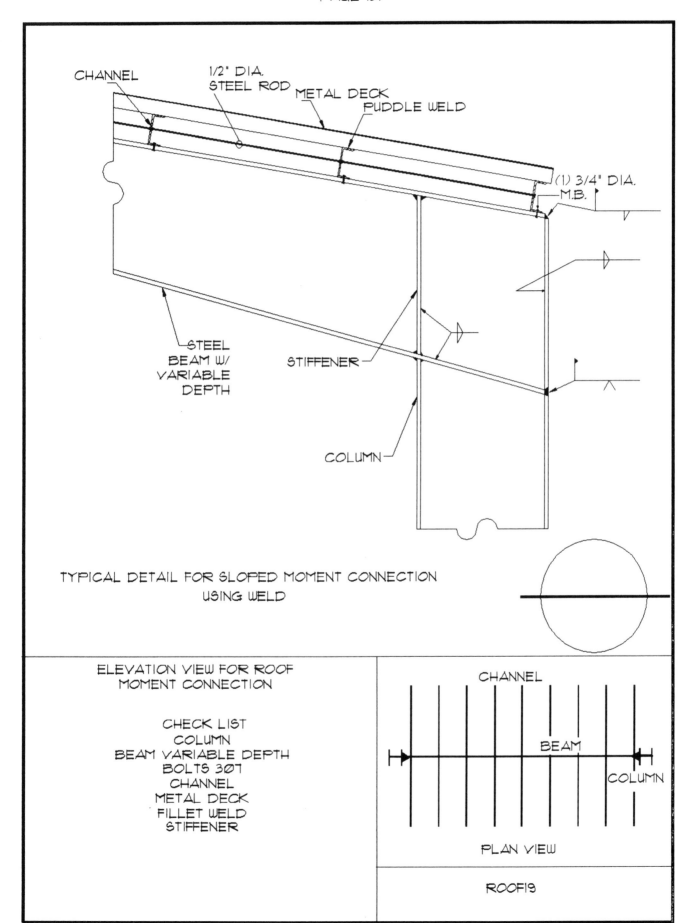

CHANNEL

1/2" DIA.
STEEL ROD

METAL DECK

PUDDLE WELD

(1) 3/4" DIA.
M.B.

STEEL
BEAM W/
VARIABLE
DEPTH

STIFFENER

COLUMN

TYPICAL DETAIL FOR SLOPED MOMENT CONNECTION
USING WELD

ELEVATION VIEW FOR ROOF
MOMENT CONNECTION

CHECK LIST
COLUMN
BEAM VARIABLE DEPTH
BOLTS 307
CHANNEL
METAL DECK
FILLET WELD
STIFFENER

CHANNEL

BEAM

COLUMN

PLAN VIEW

ROOF19

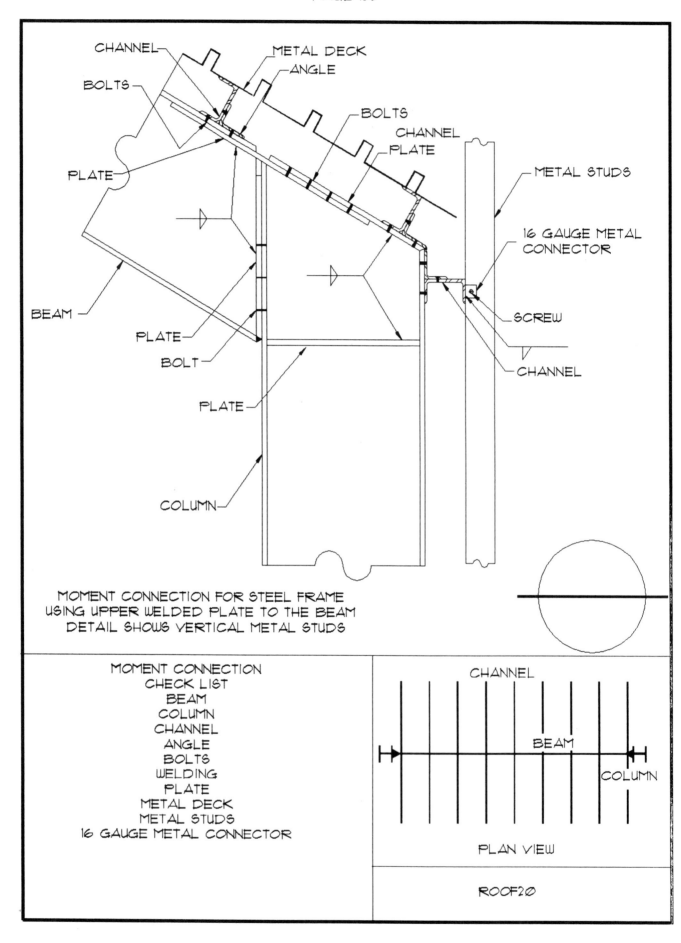

CHANNEL — METAL DECK
BOLTS — ANGLE
— BOLTS
CHANNEL
PLATE
PLATE
METAL STUDS
16 GAUGE METAL CONNECTOR
BEAM
PLATE
BOLT
SCREW
CHANNEL
PLATE
COLUMN

MOMENT CONNECTION FOR STEEL FRAME
USING UPPER WELDED PLATE TO THE BEAM
DETAIL SHOWS VERTICAL METAL STUDS

MOMENT CONNECTION
CHECK LIST
BEAM
COLUMN
CHANNEL
ANGLE
BOLTS
WELDING
PLATE
METAL DECK
METAL STUDS
16 GAUGE METAL CONNECTOR

CHANNEL
BEAM
COLUMN

PLAN VIEW

ROOF20

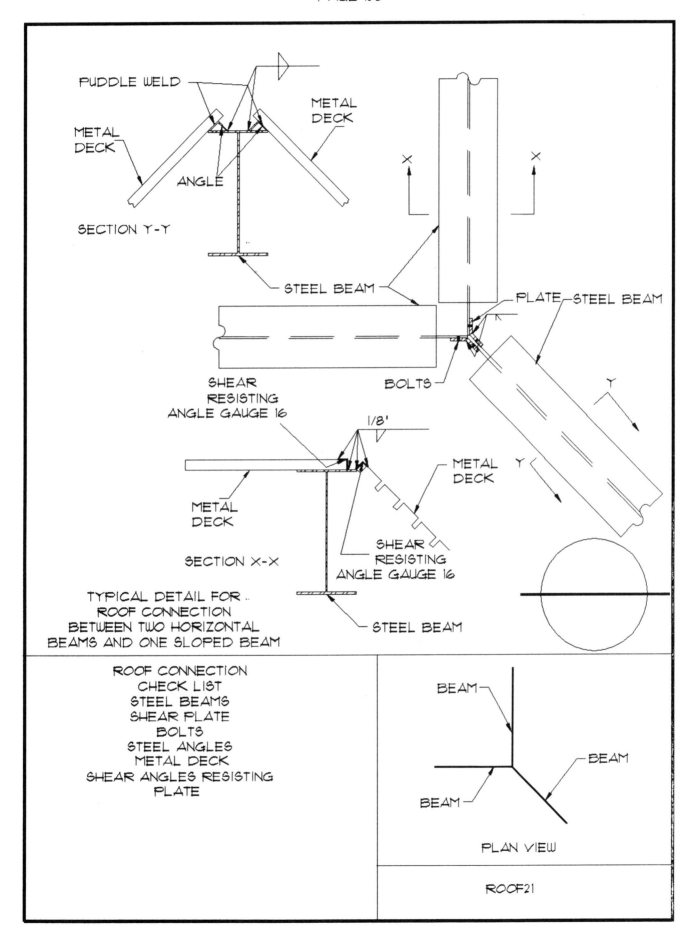

PUDDLE WELD

METAL DECK

METAL DECK

ANGLE

SECTION Y-Y

METAL DECK

STEEL BEAM

PLATE STEEL BEAM

SHEAR RESISTING ANGLE GAUGE 16

BOLTS

X X

Y

Y

1/8'

METAL DECK

METAL DECK

SHEAR RESISTING ANGLE GAUGE 16

SECTION X-X

STEEL BEAM

TYPICAL DETAIL FOR
ROOF CONNECTION
BETWEEN TWO HORIZONTAL
BEAMS AND ONE SLOPED BEAM

ROOF CONNECTION
CHECK LIST
STEEL BEAMS
SHEAR PLATE
BOLTS
STEEL ANGLES
METAL DECK
SHEAR ANGLES RESISTING
PLATE

BEAM

BEAM

BEAM

BEAM

PLAN VIEW

ROOF21

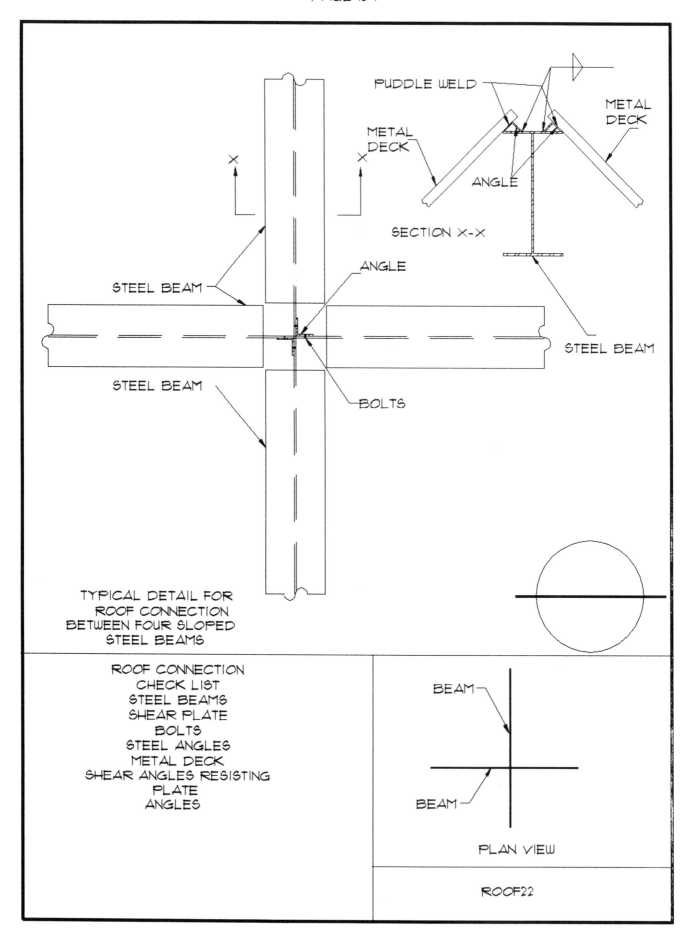

PUDDLE WELD

METAL DECK

METAL DECK

ANGLE

SECTION X-X

STEEL BEAM

ANGLE

STEEL BEAM

STEEL BEAM

BOLTS

TYPICAL DETAIL FOR
ROOF CONNECTION
BETWEEN FOUR SLOPED
STEEL BEAMS

ROOF CONNECTION
CHECK LIST
STEEL BEAMS
SHEAR PLATE
BOLTS
STEEL ANGLES
METAL DECK
SHEAR ANGLES RESISTING
PLATE
ANGLES

BEAM

BEAM

PLAN VIEW

ROOF22

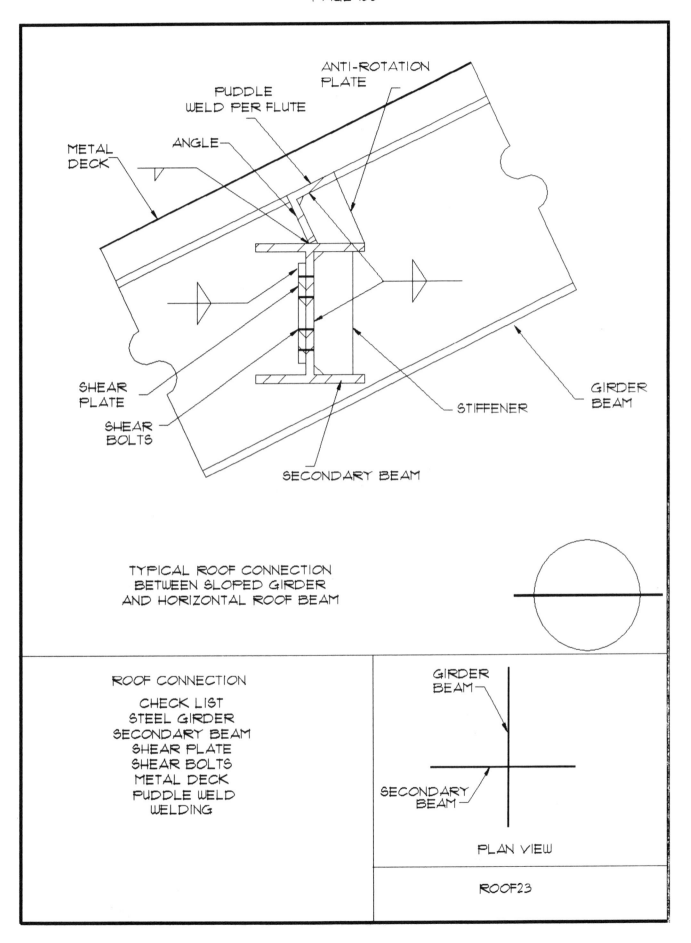

ANTI-ROTATION
PLATE

PUDDLE
WELD PER FLUTE

METAL
DECK

ANGLE

SHEAR
PLATE

SHEAR
BOLTS

STIFFENER

GIRDER
BEAM

SECONDARY BEAM

TYPICAL ROOF CONNECTION
BETWEEN SLOPED GIRDER
AND HORIZONTAL ROOF BEAM

ROOF CONNECTION

CHECK LIST
STEEL GIRDER
SECONDARY BEAM
SHEAR PLATE
SHEAR BOLTS
METAL DECK
PUDDLE WELD
WELDING

GIRDER
BEAM

SECONDARY
BEAM

PLAN VIEW

ROOF23

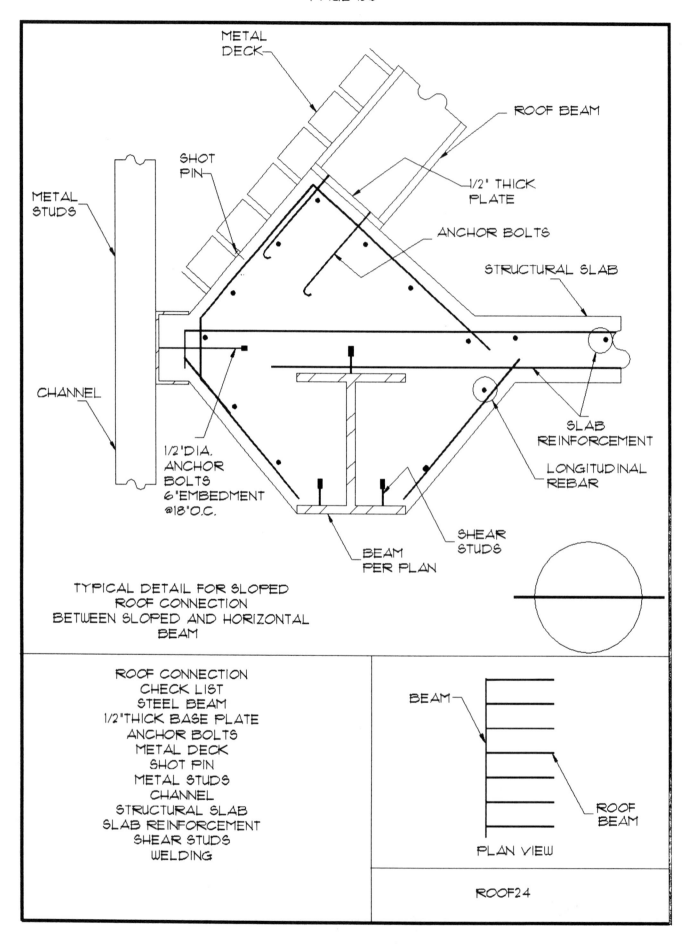

METAL DECK

ROOF BEAM

SHOT PIN

1/2" THICK PLATE

METAL STUDS

ANCHOR BOLTS

STRUCTURAL SLAB

CHANNEL

1/2'DIA. ANCHOR BOLTS 6"EMBEDMENT @18"O.C.

SLAB REINFORCEMENT

LONGITUDINAL REBAR

SHEAR STUDS

BEAM PER PLAN

TYPICAL DETAIL FOR SLOPED
ROOF CONNECTION
BETWEEN SLOPED AND HORIZONTAL
BEAM

ROOF CONNECTION
CHECK LIST
STEEL BEAM
1/2"THICK BASE PLATE
ANCHOR BOLTS
METAL DECK
SHOT PIN
METAL STUDS
CHANNEL
STRUCTURAL SLAB
SLAB REINFORCEMENT
SHEAR STUDS
WELDING

BEAM

ROOF BEAM

PLAN VIEW

ROOF24

CHAPTER 11

BUILT-UP STEEL SECTION

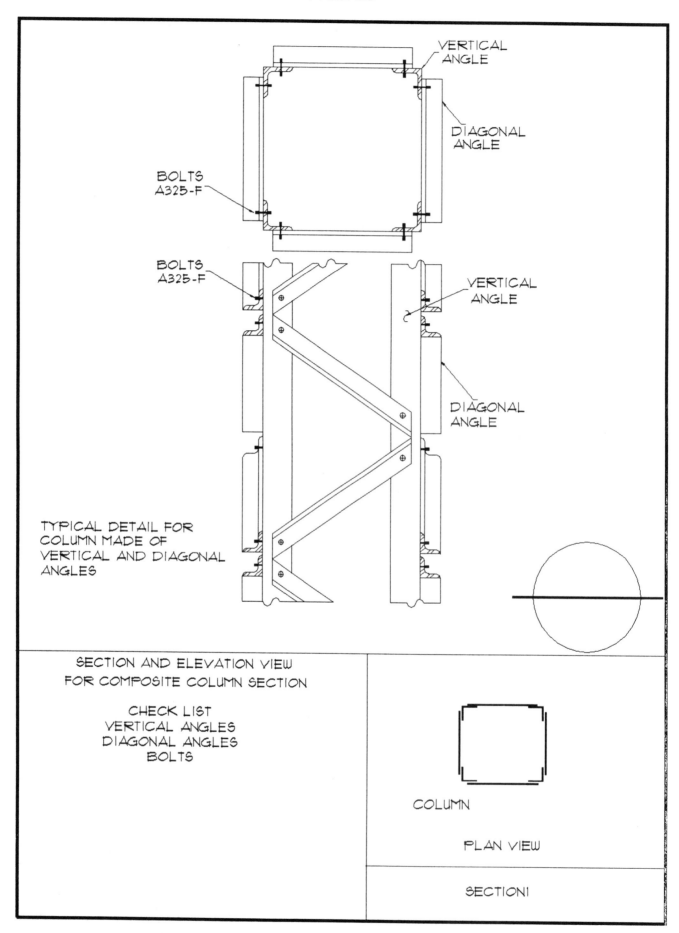

VERTICAL
ANGLE

DIAGONAL
ANGLE

BOLTS
A325-F

BOLTS
A325-F

VERTICAL
ANGLE

DIAGONAL
ANGLE

TYPICAL DETAIL FOR
COLUMN MADE OF
VERTICAL AND DIAGONAL
ANGLES

SECTION AND ELEVATION VIEW
FOR COMPOSITE COLUMN SECTION

CHECK LIST
VERTICAL ANGLES
DIAGONAL ANGLES
BOLTS

COLUMN

PLAN VIEW

SECTION1

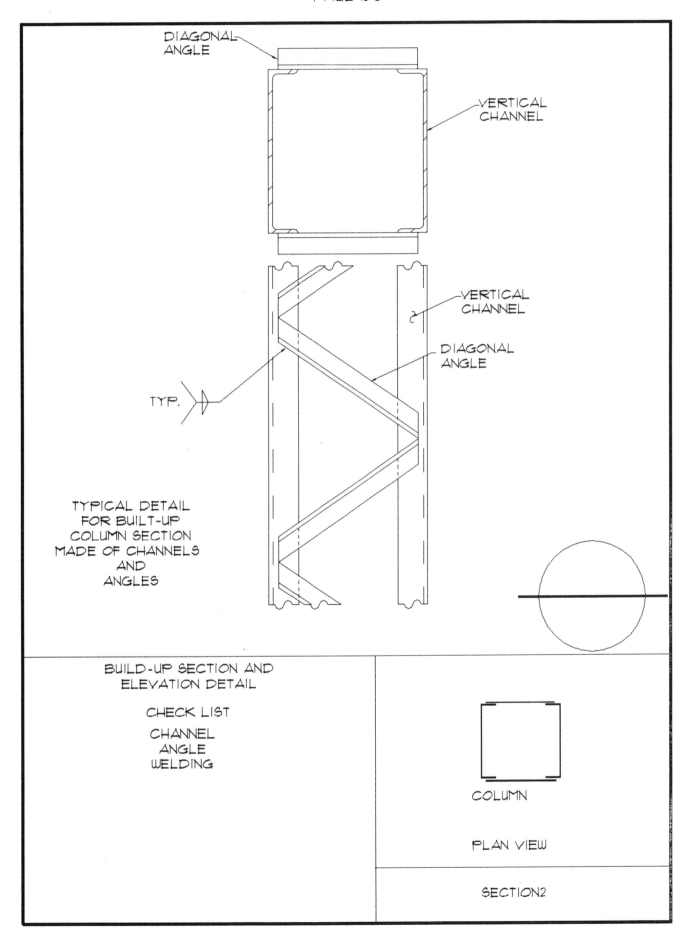

DIAGONAL
ANGLE

VERTICAL
CHANNEL

VERTICAL
CHANNEL

DIAGONAL
ANGLE

TYP.

TYPICAL DETAIL
FOR BUILT-UP
COLUMN SECTION
MADE OF CHANNELS
AND
ANGLES

BUILD-UP SECTION AND
ELEVATION DETAIL

CHECK LIST

CHANNEL
ANGLE
WELDING

COLUMN

PLAN VIEW

SECTION2

SHEAR
BOLTS

ANGLE

CHANNEL

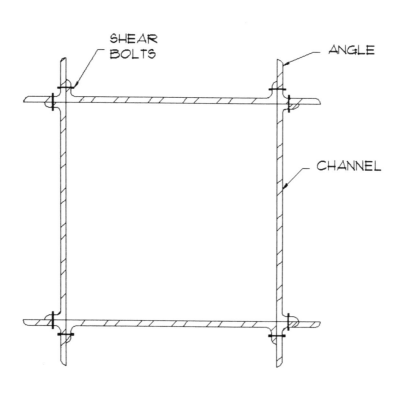

TYPICAL DETAIL FOR BUILD-UP
SECTION MADE OF BOLTED
CHANNELS & ANGLES

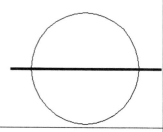

BUILD-UP SECTION AND
ELEVATION DETAIL

CHECK LIST

ANGLE
CHANNEL
BOLTS

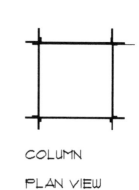

COLUMN

PLAN VIEW

SECTION3

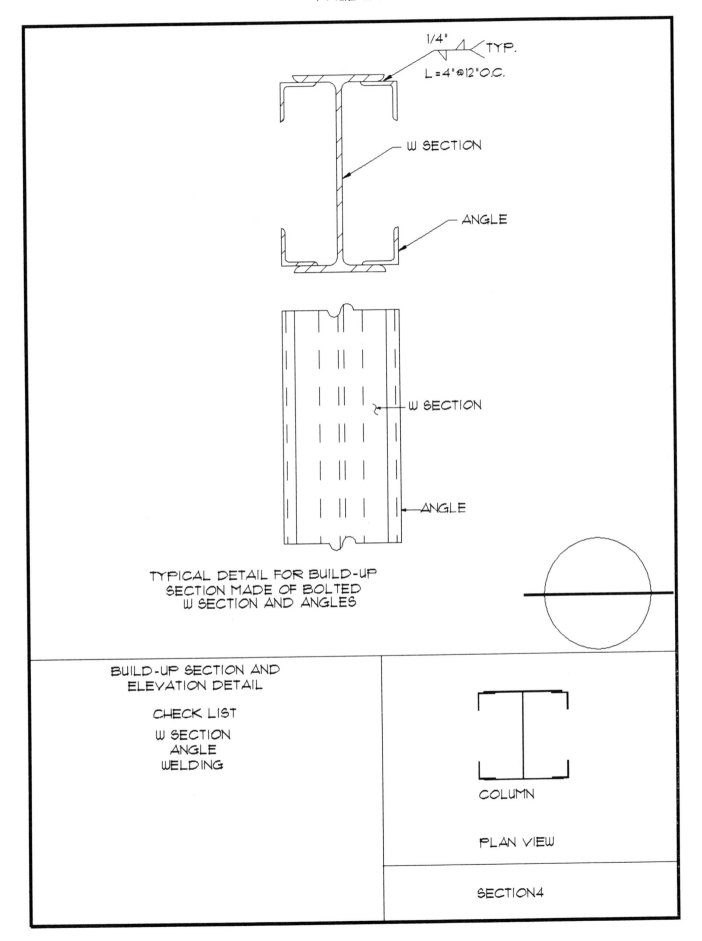

1/4" TYP.

L=4"@12"O.C.

W SECTION

ANGLE

W SECTION

ANGLE

TYPICAL DETAIL FOR BUILD-UP
SECTION MADE OF BOLTED
W SECTION AND ANGLES

BUILD-UP SECTION AND
ELEVATION DETAIL

CHECK LIST

W SECTION
ANGLE
WELDING

COLUMN

PLAN VIEW

SECTION 4

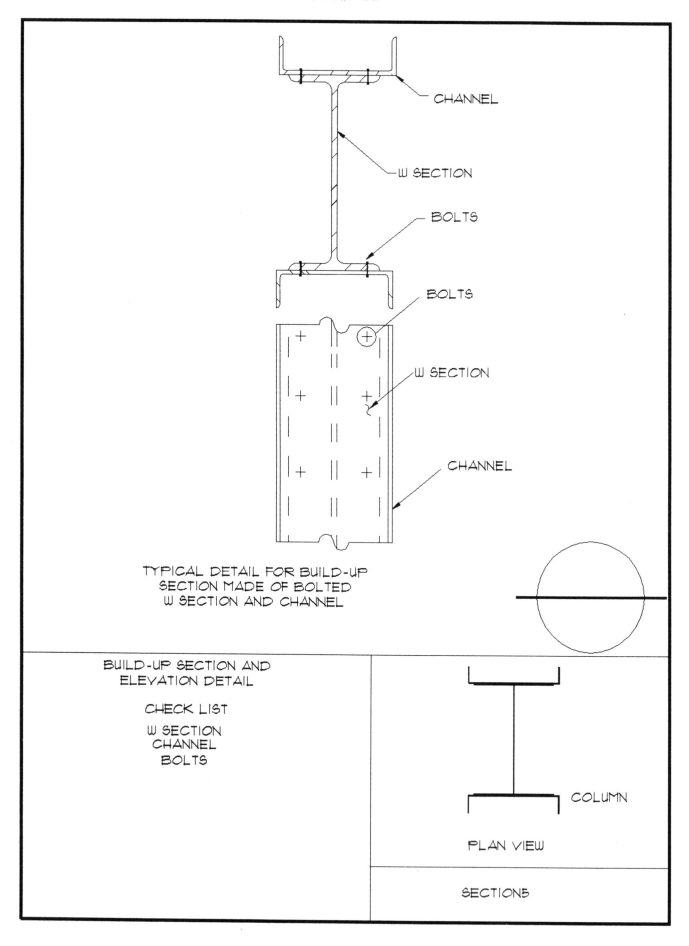

CHANNEL

W SECTION

BOLTS

BOLTS

W SECTION

CHANNEL

TYPICAL DETAIL FOR BUILD-UP
SECTION MADE OF BOLTED
W SECTION AND CHANNEL

BUILD-UP SECTION AND
ELEVATION DETAIL

CHECK LIST

W SECTION
CHANNEL
BOLTS

COLUMN

PLAN VIEW

SECTIONS

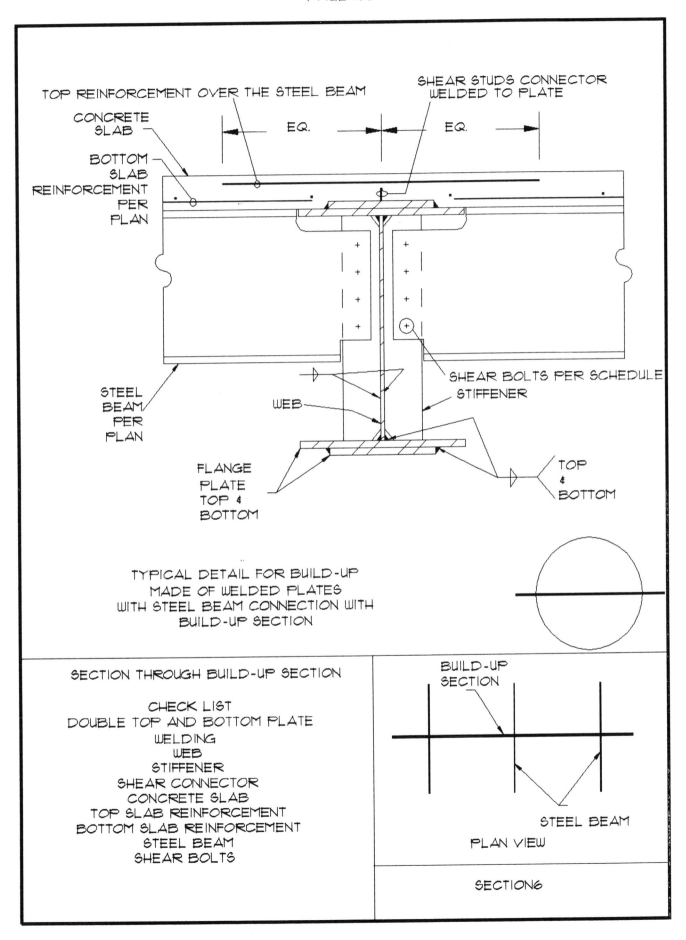

TOP REINFORCEMENT OVER THE STEEL BEAM

SHEAR STUDS CONNECTOR WELDED TO PLATE

CONCRETE SLAB

BOTTOM SLAB REINFORCEMENT PER PLAN

EQ.

EQ.

STEEL BEAM PER PLAN

WEB

SHEAR BOLTS PER SCHEDULE

STIFFENER

FLANGE PLATE TOP & BOTTOM

TOP & BOTTOM

TYPICAL DETAIL FOR BUILD-UP
MADE OF WELDED PLATES
WITH STEEL BEAM CONNECTION WITH
BUILD-UP SECTION

SECTION THROUGH BUILD-UP SECTION

CHECK LIST
DOUBLE TOP AND BOTTOM PLATE
WELDING
WEB
STIFFENER
SHEAR CONNECTOR
CONCRETE SLAB
TOP SLAB REINFORCEMENT
BOTTOM SLAB REINFORCEMENT
STEEL BEAM
SHEAR BOLTS

BUILD-UP SECTION

STEEL BEAM

PLAN VIEW

SECTION6

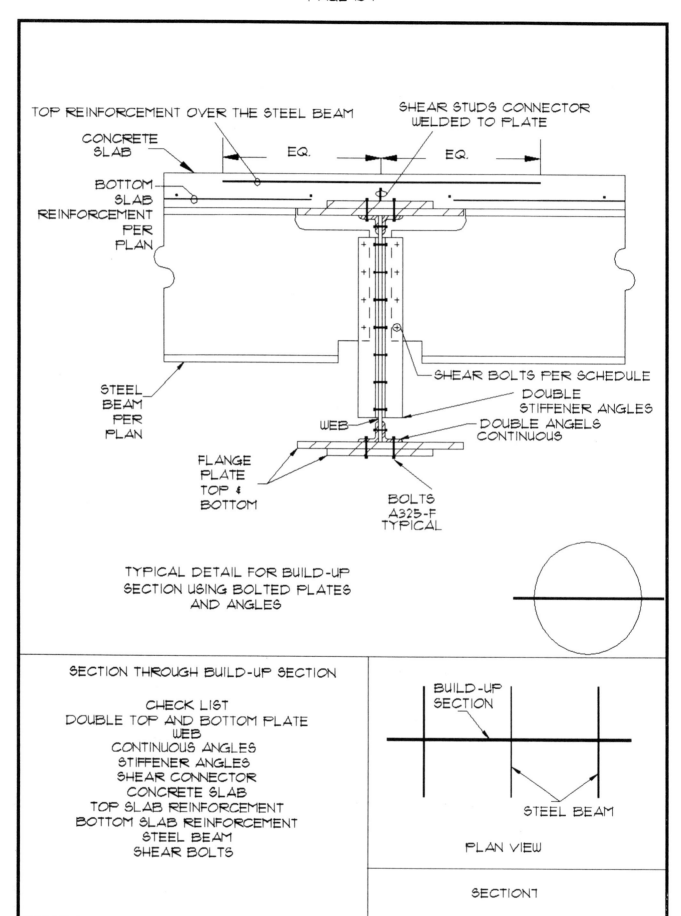

TOP REINFORCEMENT OVER THE STEEL BEAM

SHEAR STUDS CONNECTOR WELDED TO PLATE

CONCRETE SLAB

EQ.

EQ.

BOTTOM SLAB REINFORCEMENT PER PLAN

SHEAR BOLTS PER SCHEDULE

DOUBLE STIFFENER ANGLES

DOUBLE ANGELS CONTINUOUS

WEB

STEEL BEAM PER PLAN

FLANGE PLATE TOP & BOTTOM

BOLTS A325-F TYPICAL

TYPICAL DETAIL FOR BUILD-UP SECTION USING BOLTED PLATES AND ANGLES

SECTION THROUGH BUILD-UP SECTION

CHECK LIST
DOUBLE TOP AND BOTTOM PLATE
WEB
CONTINUOUS ANGLES
STIFFENER ANGLES
SHEAR CONNECTOR
CONCRETE SLAB
TOP SLAB REINFORCEMENT
BOTTOM SLAB REINFORCEMENT
STEEL BEAM
SHEAR BOLTS

BUILD-UP SECTION

STEEL BEAM

PLAN VIEW

SECTION

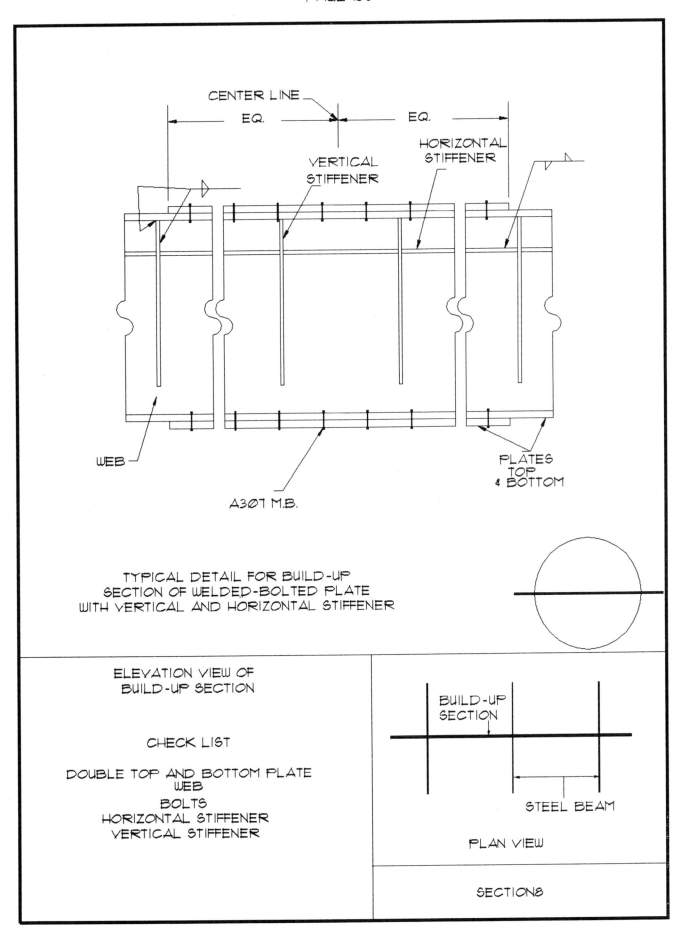

CENTER LINE

EQ.

EQ.

VERTICAL
STIFFENER

HORIZONTAL
STIFFENER

WEB

A307 M.B.

PLATES
TOP
& BOTTOM

TYPICAL DETAIL FOR BUILD-UP
SECTION OF WELDED-BOLTED PLATE
WITH VERTICAL AND HORIZONTAL STIFFENER

ELEVATION VIEW OF
BUILD-UP SECTION

CHECK LIST

DOUBLE TOP AND BOTTOM PLATE
WEB
BOLTS
HORIZONTAL STIFFENER
VERTICAL STIFFENER

BUILD-UP
SECTION

STEEL BEAM

PLAN VIEW

SECTION8

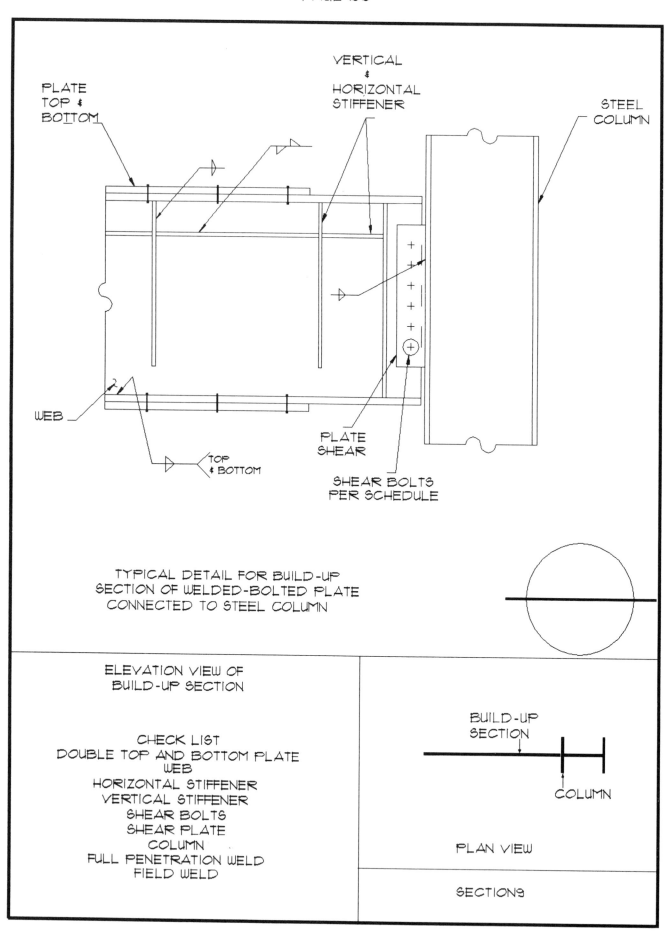

PLATE
TOP &
BOTTOM

VERTICAL
&
HORIZONTAL
STIFFENER

STEEL
COLUMN

WEB

PLATE
SHEAR

TOP
& BOTTOM

SHEAR BOLTS
PER SCHEDULE

TYPICAL DETAIL FOR BUILD-UP
SECTION OF WELDED-BOLTED PLATE
CONNECTED TO STEEL COLUMN

ELEVATION VIEW OF
BUILD-UP SECTION

CHECK LIST
DOUBLE TOP AND BOTTOM PLATE
WEB
HORIZONTAL STIFFENER
VERTICAL STIFFENER
SHEAR BOLTS
SHEAR PLATE
COLUMN
FULL PENETRATION WELD
FIELD WELD

BUILD-UP
SECTION

COLUMN

PLAN VIEW

SECTION9

CHAPTER 12

STEEL SHEAR CONNECTION

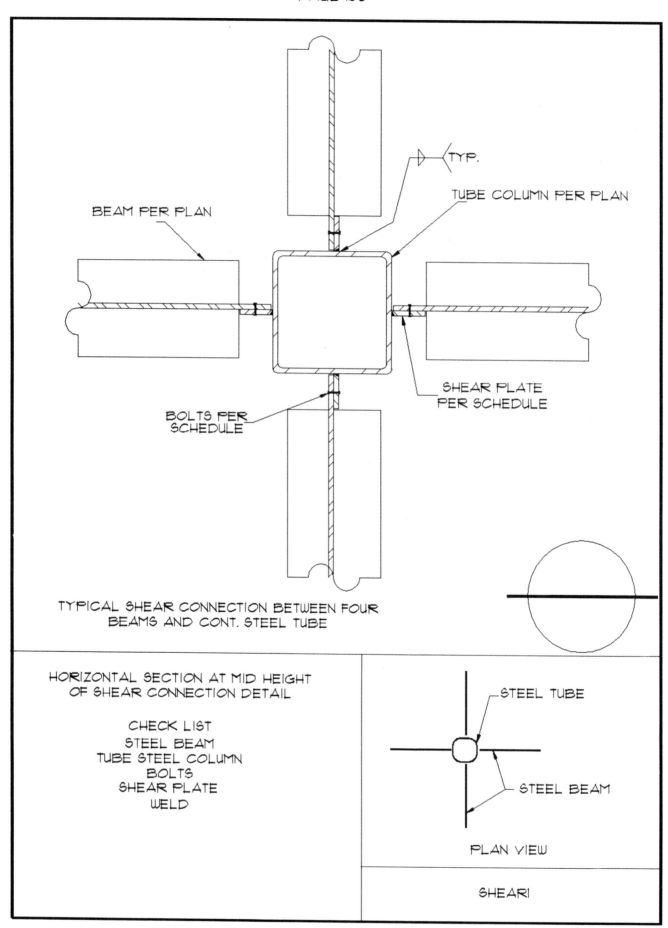

TYP.

BEAM PER PLAN

TUBE COLUMN PER PLAN

SHEAR PLATE
PER SCHEDULE

BOLTS PER
SCHEDULE

TYPICAL SHEAR CONNECTION BETWEEN FOUR
BEAMS AND CONT. STEEL TUBE

HORIZONTAL SECTION AT MID HEIGHT
OF SHEAR CONNECTION DETAIL

CHECK LIST
STEEL BEAM
TUBE STEEL COLUMN
BOLTS
SHEAR PLATE
WELD

STEEL TUBE

STEEL BEAM

PLAN VIEW

SHEAR1

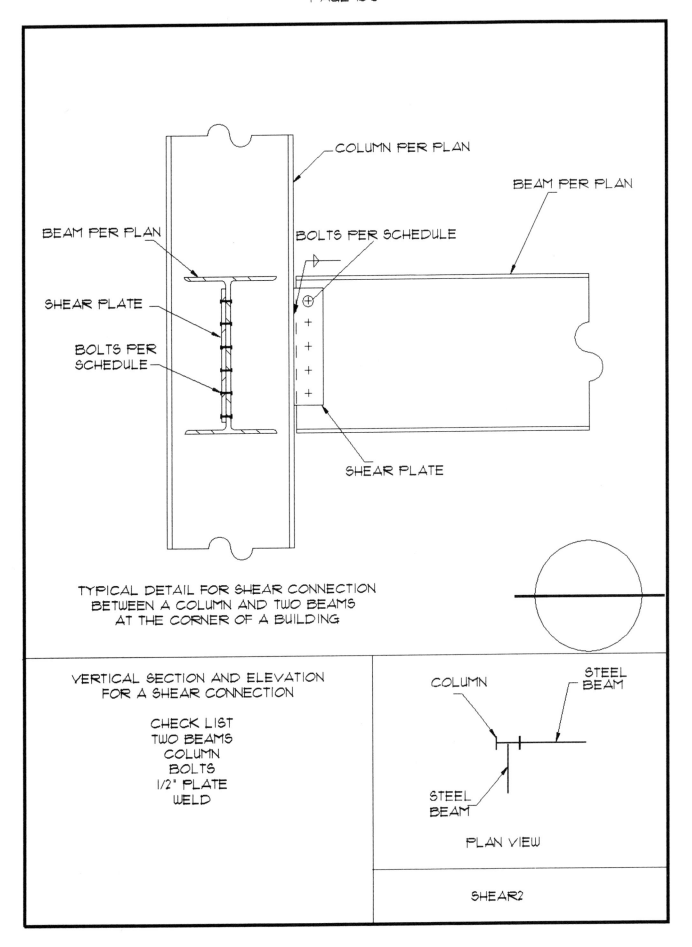

TYPICAL DETAIL FOR SHEAR CONNECTION
BETWEEN A COLUMN AND TWO BEAMS
AT THE CORNER OF A BUILDING

VERTICAL SECTION AND ELEVATION
FOR A SHEAR CONNECTION

CHECK LIST
TWO BEAMS
COLUMN
BOLTS
1/2" PLATE
WELD

PLAN VIEW

SHEAR2

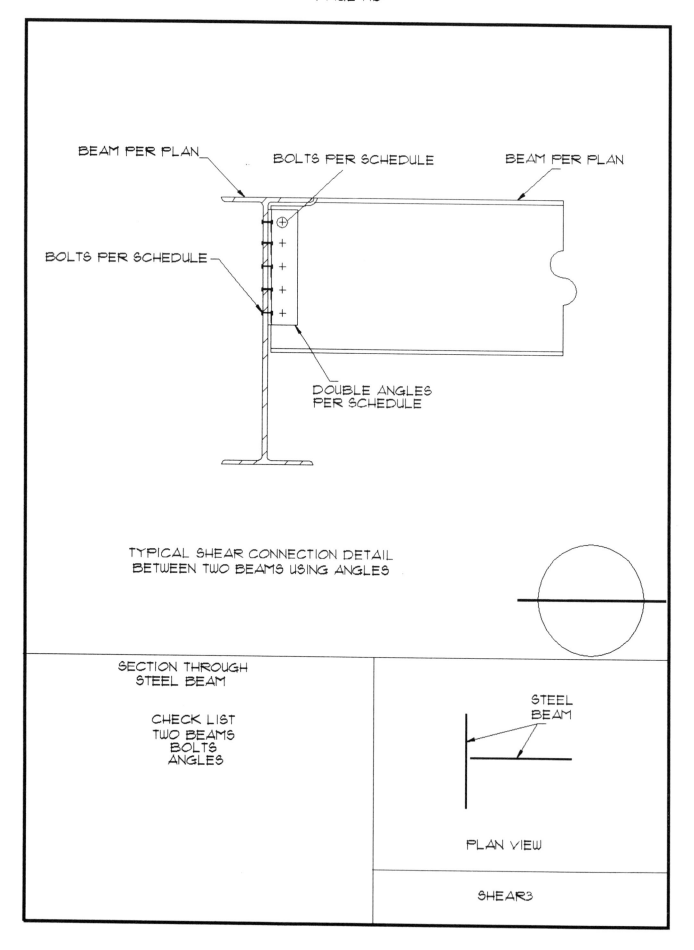

BEAM PER PLAN

BOLTS PER SCHEDULE

BEAM PER PLAN

BOLTS PER SCHEDULE

DOUBLE ANGLES
PER SCHEDULE

TYPICAL SHEAR CONNECTION DETAIL
BETWEEN TWO BEAMS USING ANGLES

SECTION THROUGH
STEEL BEAM

CHECK LIST
TWO BEAMS
BOLTS
ANGLES

STEEL
BEAM

PLAN VIEW

SHEAR3

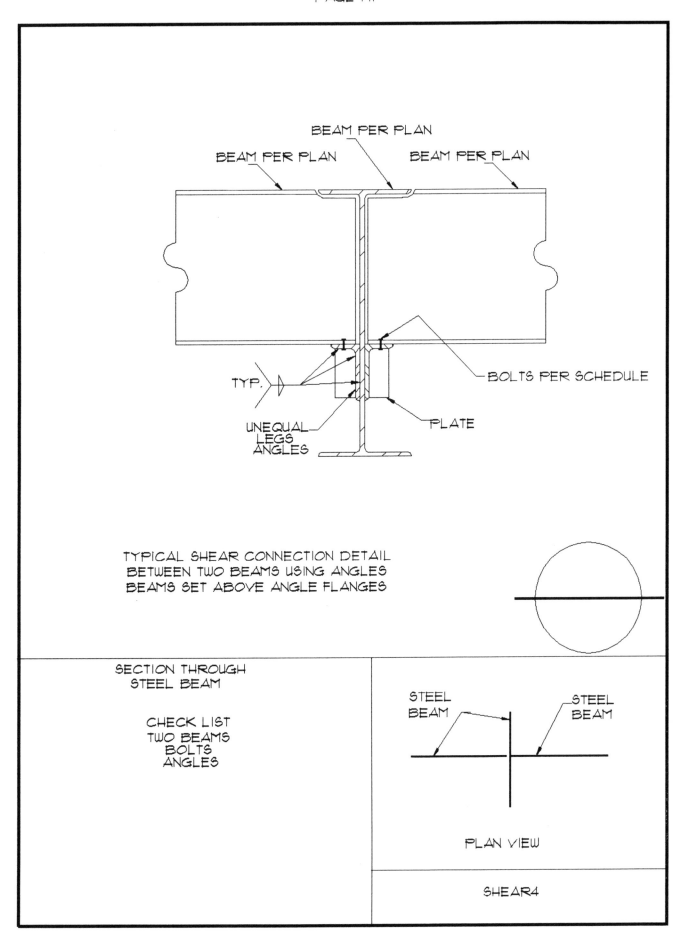

BEAM PER PLAN

BEAM PER PLAN

BEAM PER PLAN

BOLTS PER SCHEDULE

TYP.

UNEQUAL
LEGS
ANGLES

PLATE

TYPICAL SHEAR CONNECTION DETAIL
BETWEEN TWO BEAMS USING ANGLES
BEAMS SET ABOVE ANGLE FLANGES

SECTION THROUGH
STEEL BEAM

CHECK LIST
TWO BEAMS
BOLTS
ANGLES

STEEL
BEAM

STEEL
BEAM

PLAN VIEW

SHEAR4

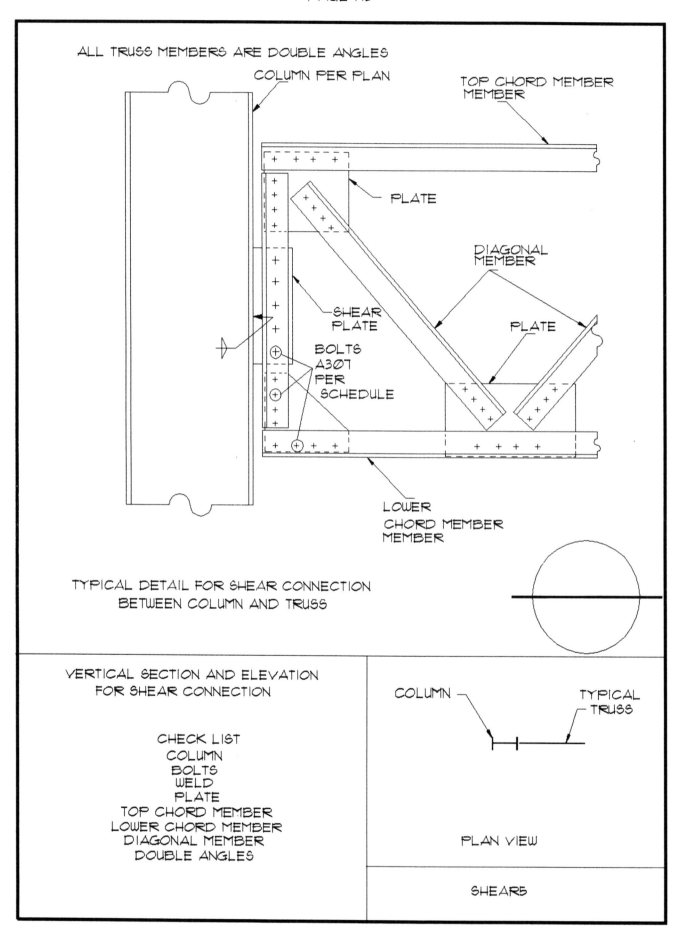

ALL TRUSS MEMBERS ARE DOUBLE ANGLES

COLUMN PER PLAN

TOP CHORD MEMBER MEMBER

PLATE

DIAGONAL MEMBER

PLATE

SHEAR PLATE

BOLTS A307 PER SCHEDULE

LOWER CHORD MEMBER MEMBER

TYPICAL DETAIL FOR SHEAR CONNECTION
BETWEEN COLUMN AND TRUSS

VERTICAL SECTION AND ELEVATION
FOR SHEAR CONNECTION

CHECK LIST
COLUMN
BOLTS
WELD
PLATE
TOP CHORD MEMBER
LOWER CHORD MEMBER
DIAGONAL MEMBER
DOUBLE ANGLES

COLUMN

TYPICAL TRUSS

PLAN VIEW

SHEARS

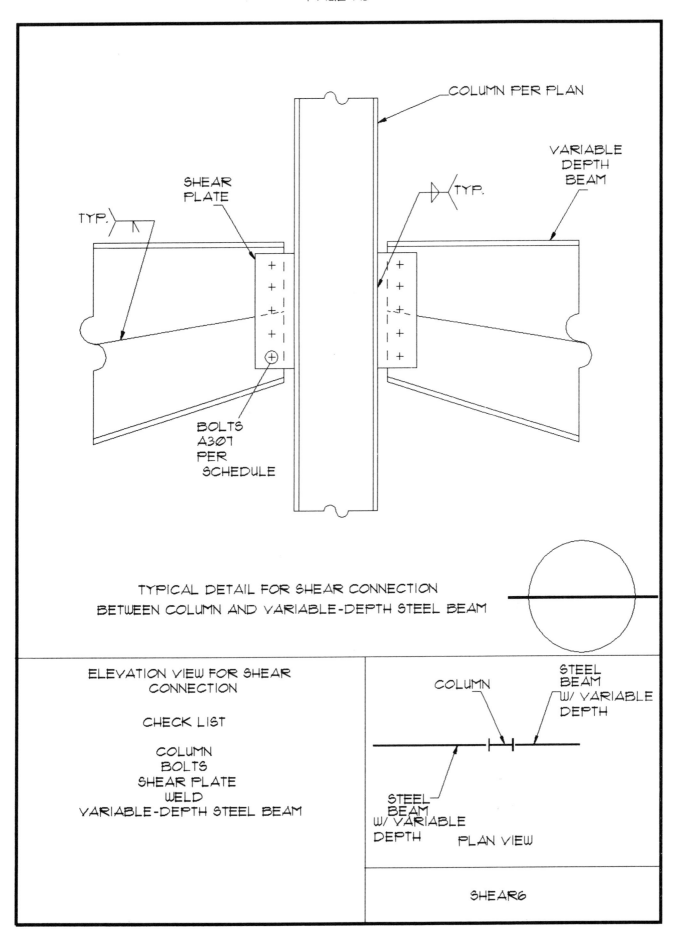

COLUMN PER PLAN

VARIABLE DEPTH BEAM

SHEAR PLATE

TYP.

TYP.

BOLTS A307 PER SCHEDULE

TYPICAL DETAIL FOR SHEAR CONNECTION
BETWEEN COLUMN AND VARIABLE-DEPTH STEEL BEAM

ELEVATION VIEW FOR SHEAR CONNECTION

CHECK LIST

COLUMN
BOLTS
SHEAR PLATE
WELD
VARIABLE-DEPTH STEEL BEAM

STEEL BEAM W/ VARIABLE DEPTH

COLUMN

STEEL BEAM W/ VARIABLE DEPTH

PLAN VIEW

SHEAR6

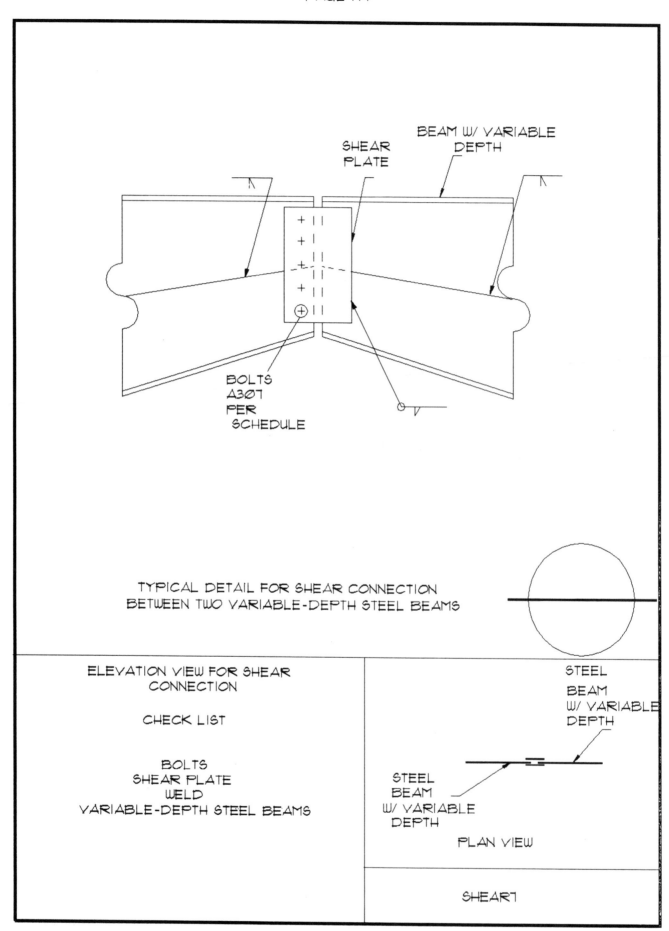

BEAM W/ VARIABLE DEPTH

SHEAR PLATE

BOLTS
A307
PER
SCHEDULE

TYPICAL DETAIL FOR SHEAR CONNECTION
BETWEEN TWO VARIABLE-DEPTH STEEL BEAMS

ELEVATION VIEW FOR SHEAR
CONNECTION

CHECK LIST

BOLTS
SHEAR PLATE
WELD
VARIABLE-DEPTH STEEL BEAMS

STEEL
BEAM
W/ VARIABLE
DEPTH

STEEL
BEAM
W/ VARIABLE
DEPTH

PLAN VIEW

SHEAR

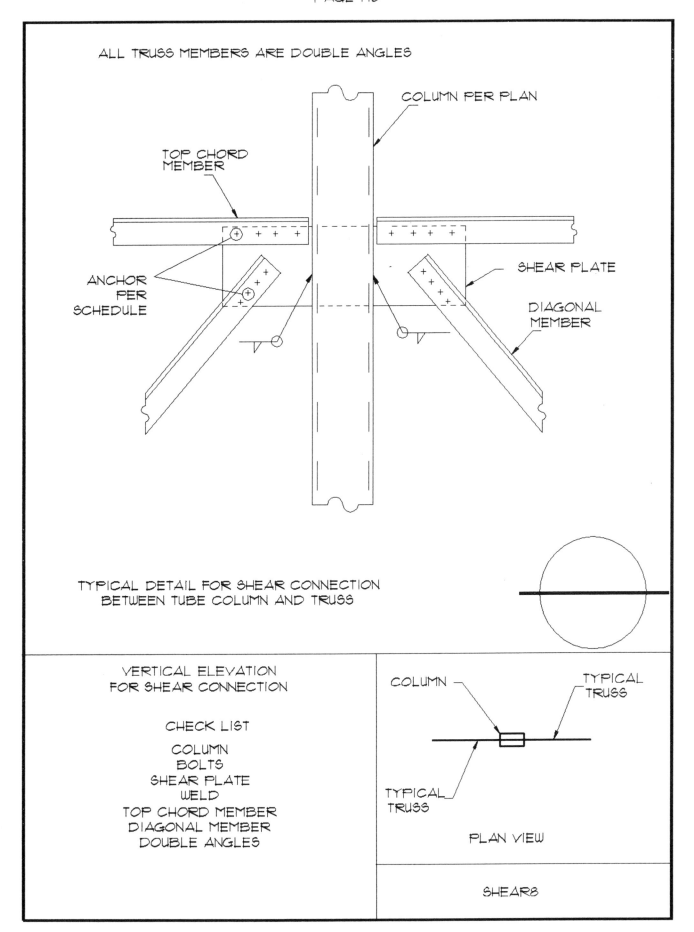

ALL TRUSS MEMBERS ARE DOUBLE ANGLES

COLUMN PER PLAN

TOP CHORD
MEMBER

ANCHOR
PER
SCHEDULE

SHEAR PLATE

DIAGONAL
MEMBER

TYPICAL DETAIL FOR SHEAR CONNECTION
BETWEEN TUBE COLUMN AND TRUSS

VERTICAL ELEVATION
FOR SHEAR CONNECTION

CHECK LIST

COLUMN
BOLTS
SHEAR PLATE
WELD
TOP CHORD MEMBER
DIAGONAL MEMBER
DOUBLE ANGLES

COLUMN

TYPICAL
TRUSS

TYPICAL
TRUSS

PLAN VIEW

SHEAR8

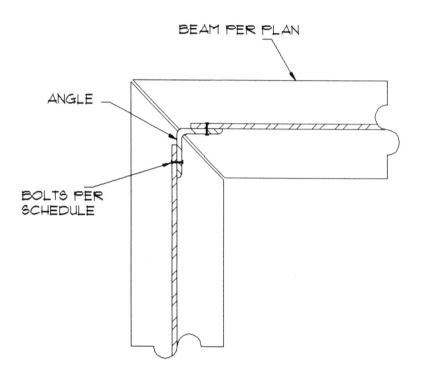

BEAM PER PLAN

ANGLE

BOLTS PER
SCHEDULE

TYPICAL SHEAR CONNECTION AT
THE INTERSECTION OF TWO
STEEL BEAMS

HORIZONTAL SECTION FOR
SHEAR CONNECTION

CHECK LIST

STEEL BEAMS
ANGLES
BOLTS

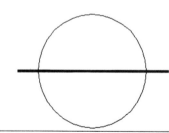

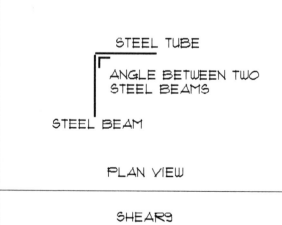

STEEL TUBE

ANGLE BETWEEN TWO
STEEL BEAMS

STEEL BEAM

PLAN VIEW

SHEAR9

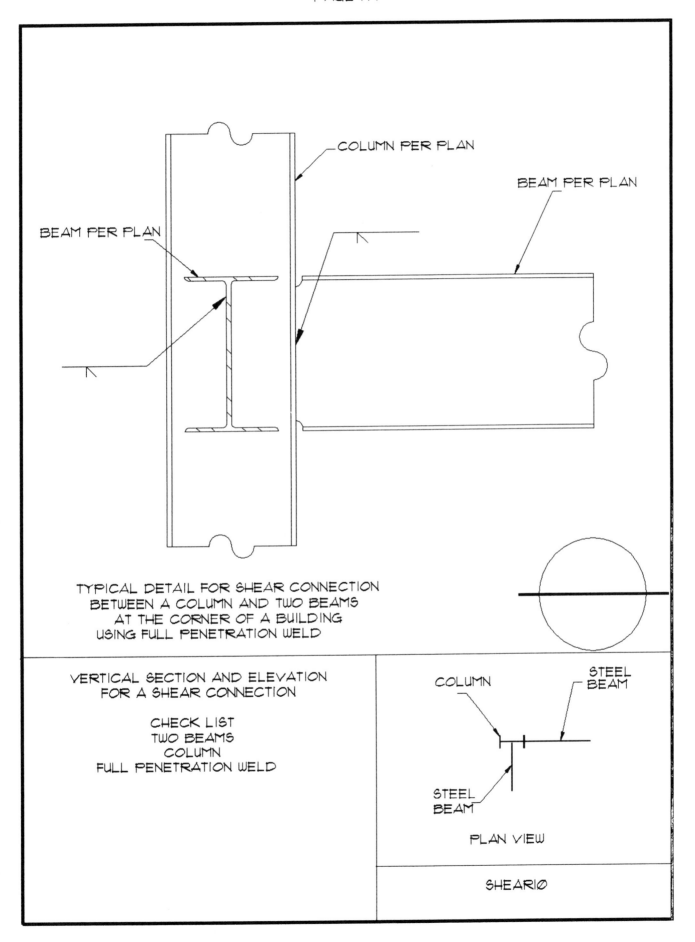

COLUMN PER PLAN

BEAM PER PLAN

BEAM PER PLAN

TYPICAL DETAIL FOR SHEAR CONNECTION
BETWEEN A COLUMN AND TWO BEAMS
AT THE CORNER OF A BUILDING
USING FULL PENETRATION WELD

VERTICAL SECTION AND ELEVATION
FOR A SHEAR CONNECTION

CHECK LIST
TWO BEAMS
COLUMN
FULL PENETRATION WELD

COLUMN

STEEL
BEAM

STEEL
BEAM

PLAN VIEW

SHEAR10

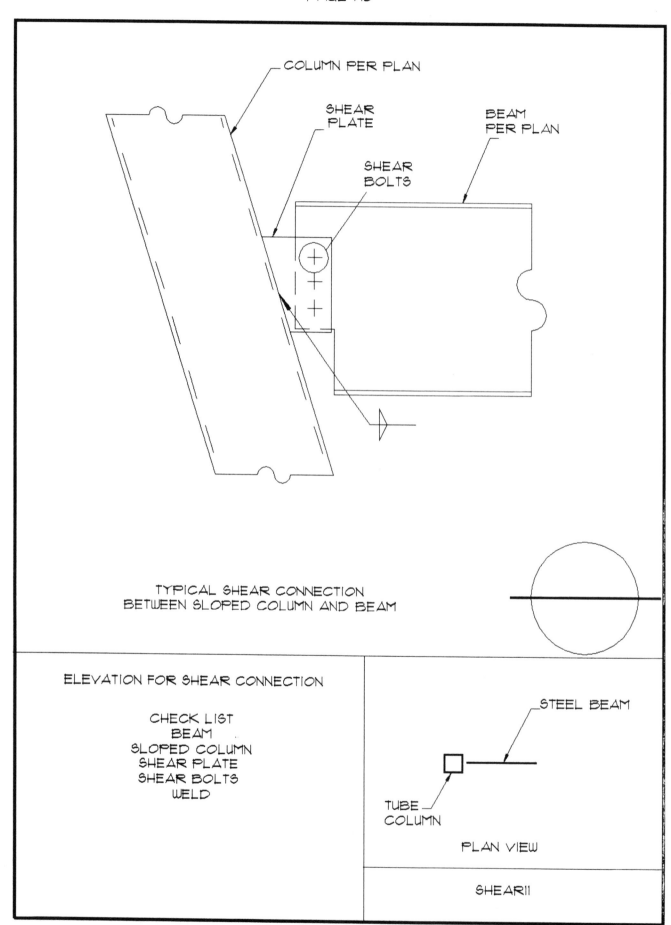

COLUMN PER PLAN

SHEAR PLATE

BEAM PER PLAN

SHEAR BOLTS

TYPICAL SHEAR CONNECTION
BETWEEN SLOPED COLUMN AND BEAM

ELEVATION FOR SHEAR CONNECTION

CHECK LIST
BEAM
SLOPED COLUMN
SHEAR PLATE
SHEAR BOLTS
WELD

STEEL BEAM

TUBE COLUMN

PLAN VIEW

SHEAR11

SHEAR CONNECTION SCHEDULE

BEAM	PLATE	BOLTS	WELD
W12	1/4 "	(4)3/8" A307 M.B.	3/16"
W18	3/8"	(5)3/4" A307 M.B.	3/16"
W24	1/2"	(6)7/8" A307 M.B.	1/4'
W27	1/2"	(6)1" A307 M.B.	1/4'
W36	3/4"	(6) 1" A307 M.B.	1/4"

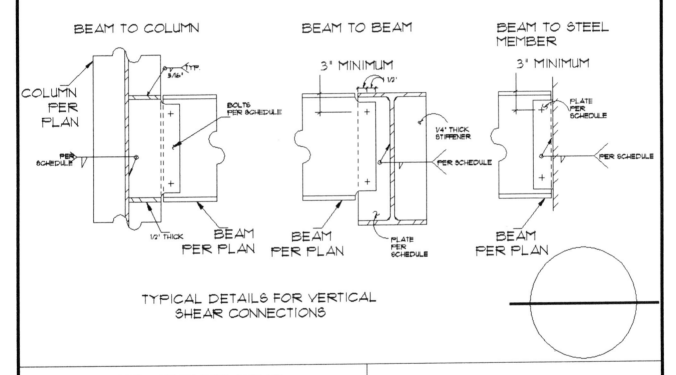

BEAM TO COLUMN

COLUMN PER PLAN

PER SCHEDULE

3/16' TYP.

BOLTS PER SCHEDULE

1/2' THICK

BEAM PER PLAN

BEAM TO BEAM

3" MINIMUM

1/2'

1/4' THICK STIFFENER

PER SCHEDULE

BEAM PER PLAN

PLATE PER SCHEDULE

BEAM TO STEEL MEMBER

3" MINIMUM

PLATE PER SCHEDULE

PER SCHEDULE

BEAM PER PLAN

TYPICAL DETAILS FOR VERTICAL
SHEAR CONNECTIONS

ELEVATION FOR SHEAR
CONNECTORS

CHECK LIST

COLUMN
BEAM
BOLTS
WELD
STIFFENER
PLATE
STEEL SURFACE

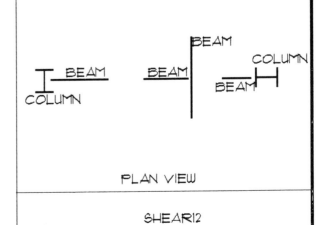

PLAN VIEW

SHEAR12

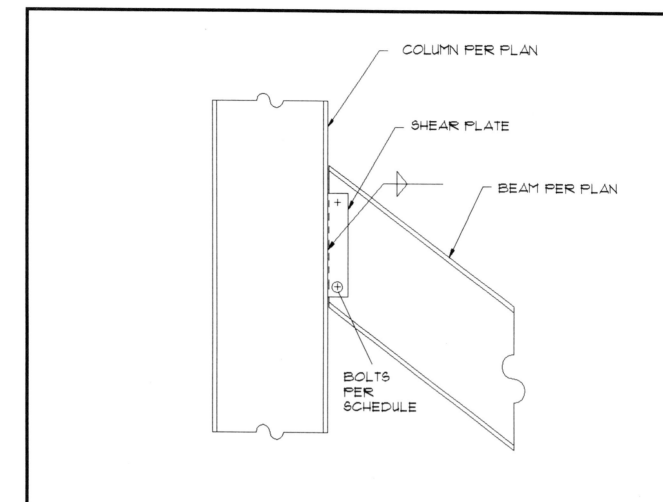

COLUMN PER PLAN

SHEAR PLATE

BEAM PER PLAN

BOLTS
PER
SCHEDULE

TYPICAL DETAIL FOR SHEAR CONNECTION
BETWEEN CONTINUOUS COLUMN AND SLOPED BEAM

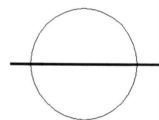

TYPICAL DETAIL FOR SHEAR CONNECTION

CHECK LIST
BEAM
COLUMN
SHEAR PLATE
WELD
BOLTS

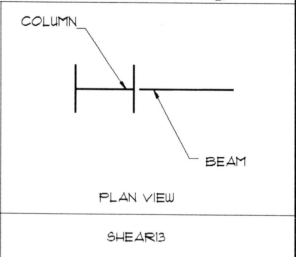

COLUMN

BEAM

PLAN VIEW

SHEAR13

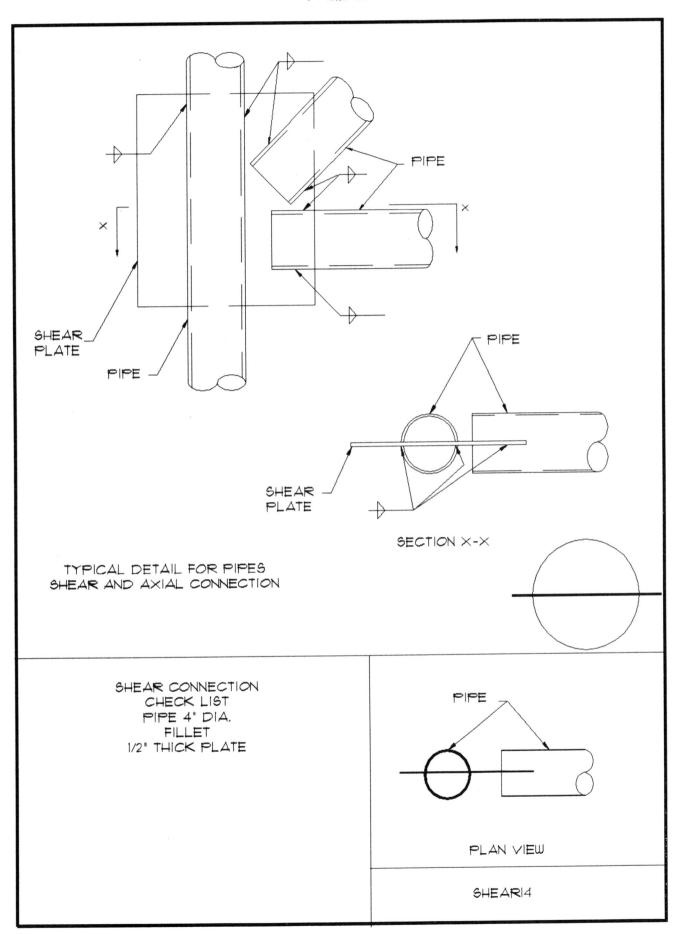

PIPE

X

SHEAR
PLATE

PIPE

PIPE

SECTION X-X

TYPICAL DETAIL FOR PIPES
SHEAR AND AXIAL CONNECTION

SHEAR
PLATE

SHEAR CONNECTION
CHECK LIST
PIPE 4" DIA.
FILLET
1/2" THICK PLATE

PIPE

PLAN VIEW

SHEAR14

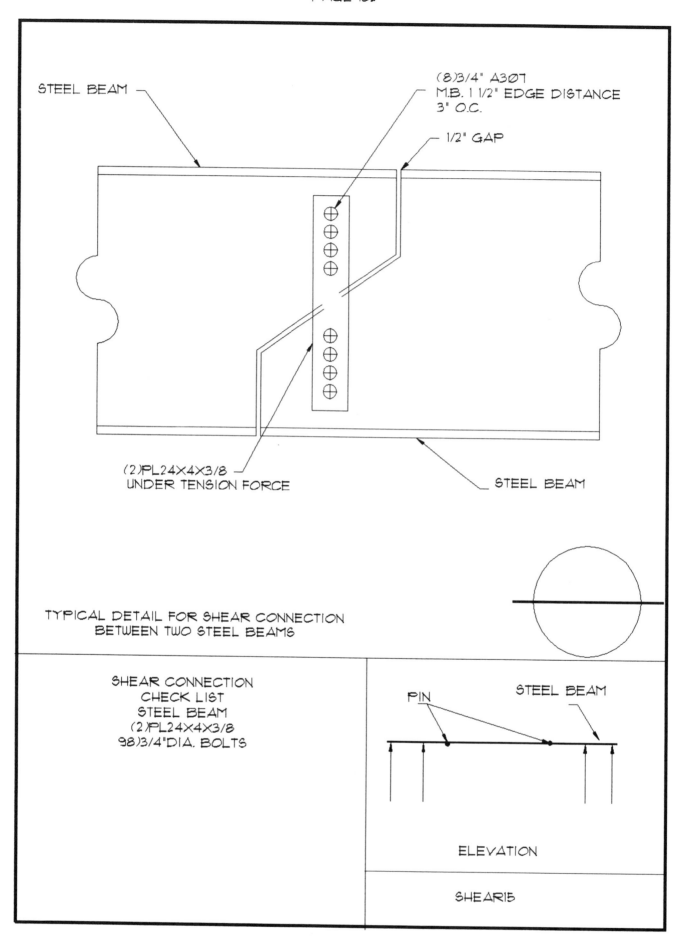

STEEL BEAM

(8)3/4" A307
M.B. 1 1/2" EDGE DISTANCE
3" O.C.

1/2" GAP

(2)PL24X4X3/8
UNDER TENSION FORCE

STEEL BEAM

TYPICAL DETAIL FOR SHEAR CONNECTION
BETWEEN TWO STEEL BEAMS

SHEAR CONNECTION
CHECK LIST
STEEL BEAM
(2)PL24X4X3/8
98)3/4"DIA. BOLTS

PIN   STEEL BEAM

ELEVATION

SHEAR15

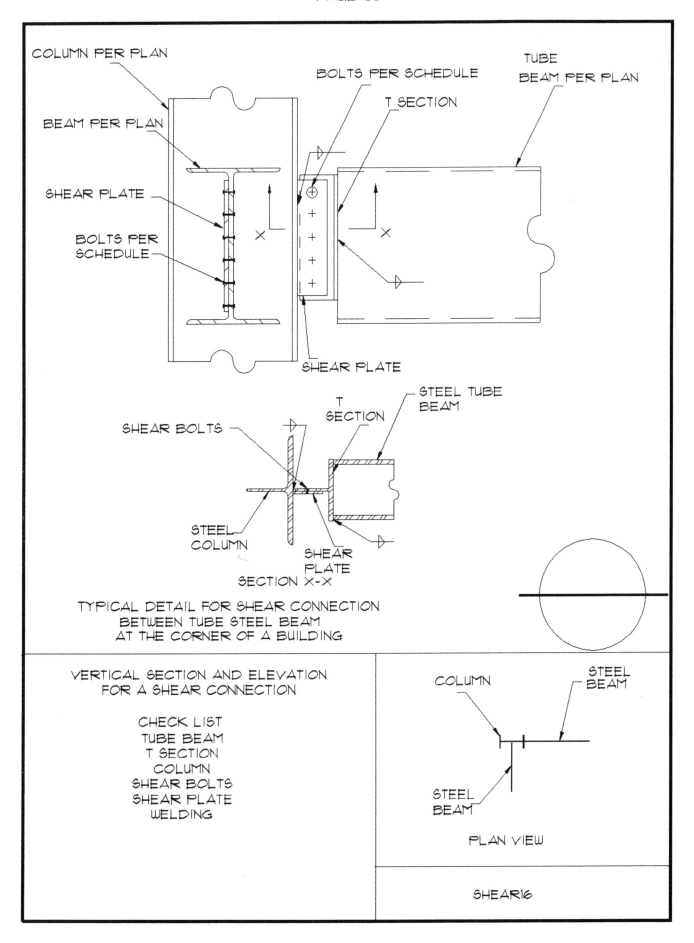

COLUMN PER PLAN

BOLTS PER SCHEDULE

T SECTION

TUBE
BEAM PER PLAN

BEAM PER PLAN

SHEAR PLATE

BOLTS PER
SCHEDULE

SHEAR PLATE

SHEAR BOLTS

T SECTION

STEEL TUBE
BEAM

STEEL
COLUMN

SHEAR
PLATE

SECTION X-X

TYPICAL DETAIL FOR SHEAR CONNECTION
BETWEEN TUBE STEEL BEAM
AT THE CORNER OF A BUILDING

VERTICAL SECTION AND ELEVATION
FOR A SHEAR CONNECTION

CHECK LIST
TUBE BEAM
T SECTION
COLUMN
SHEAR BOLTS
SHEAR PLATE
WELDING

COLUMN

STEEL
BEAM

STEEL
BEAM

PLAN VIEW

SHEAR16

CHAPTER 13

CONNECTION WITH LONGITUDINAL SLIDING

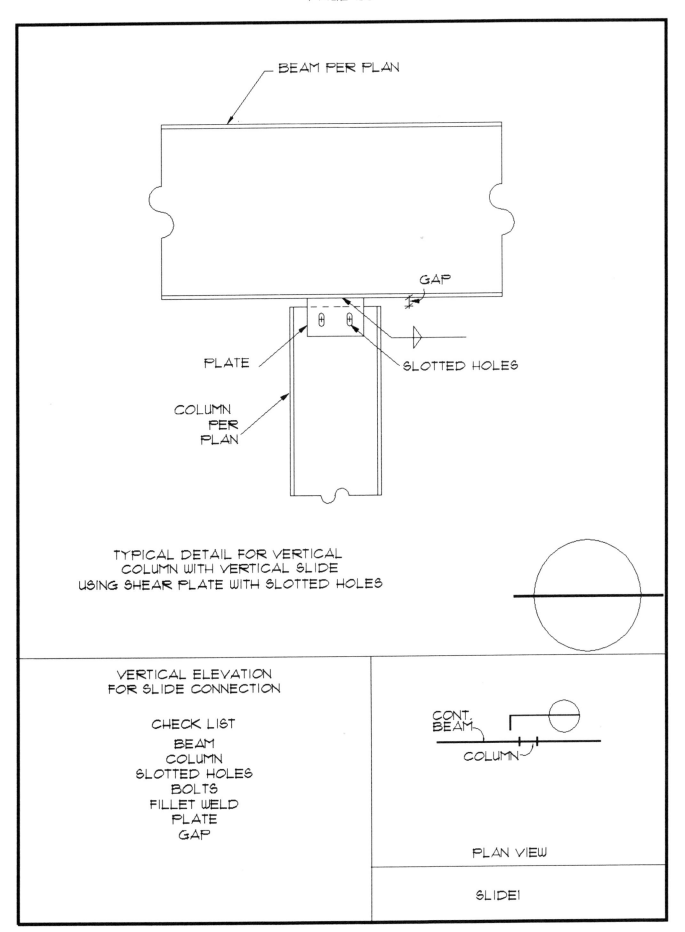

BEAM PER PLAN

GAP

PLATE

SLOTTED HOLES

COLUMN
PER
PLAN

TYPICAL DETAIL FOR VERTICAL
COLUMN WITH VERTICAL SLIDE
USING SHEAR PLATE WITH SLOTTED HOLES

VERTICAL ELEVATION
FOR SLIDE CONNECTION

CHECK LIST

BEAM
COLUMN
SLOTTED HOLES
BOLTS
FILLET WELD
PLATE
GAP

CONT.
BEAM

COLUMN

PLAN VIEW

SLIDE1

PAGE 186

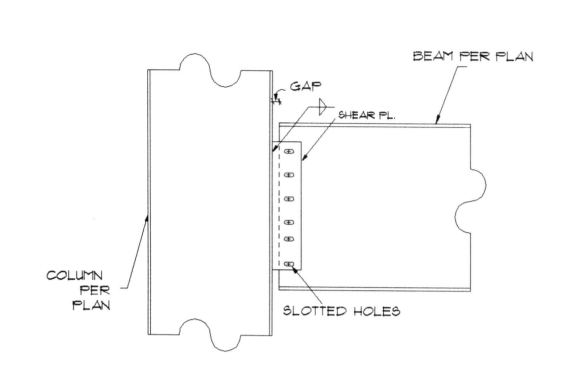

BEAM PER PLAN

GAP

SHEAR PL.

COLUMN
PER
PLAN

SLOTTED HOLES

TYPICAL DETAIL FOR HORIZONTAL BEAM ABLE TO
SLIDE HORIZONTALLY AGAINST COLUMN

VERTICAL ELEVATION
FOR SLIDE CONNECTION

CHECK LIST
BEAM
COLUMN
SLOTTED HOLES
BOLTS
FILLET WELD
PLATE
GAP

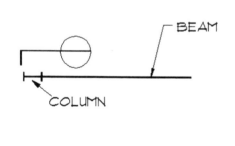

BEAM

COLUMN

PLAN VIEW

SLIDE2

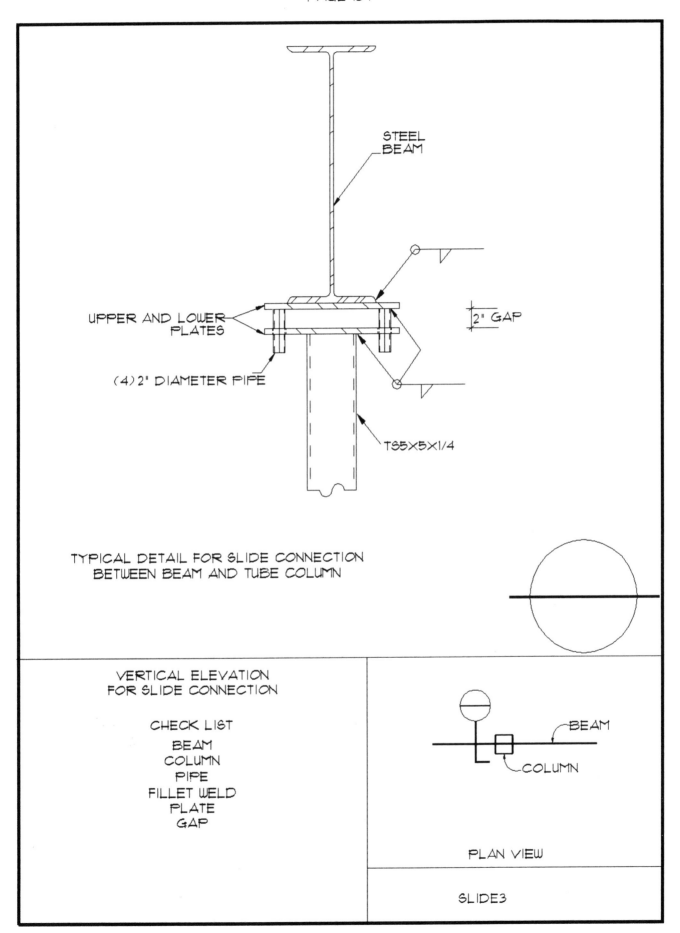

STEEL
BEAM

UPPER AND LOWER
PLATES

2" GAP

(4) 2" DIAMETER PIPE

TS5X5X1/4

TYPICAL DETAIL FOR SLIDE CONNECTION
BETWEEN BEAM AND TUBE COLUMN

VERTICAL ELEVATION
FOR SLIDE CONNECTION

CHECK LIST
BEAM
COLUMN
PIPE
FILLET WELD
PLATE
GAP

BEAM

COLUMN

PLAN VIEW

SLIDE3

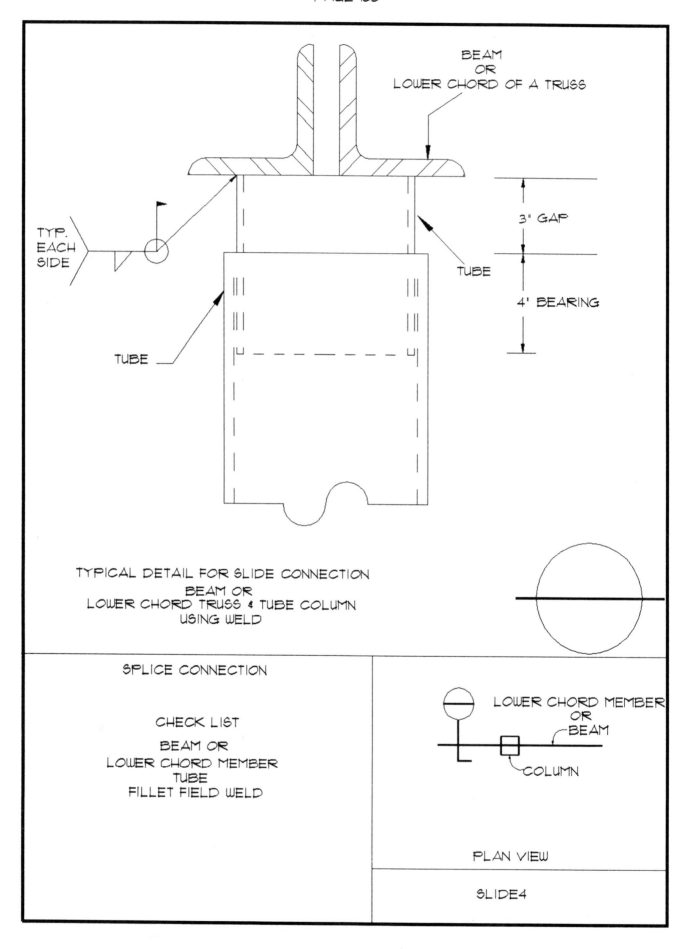

TYP.
EACH
SIDE

BEAM
OR
LOWER CHORD OF A TRUSS

3" GAP

4' BEARING

TUBE

TUBE

TYPICAL DETAIL FOR SLIDE CONNECTION
BEAM OR
LOWER CHORD TRUSS & TUBE COLUMN
USING WELD

SPLICE CONNECTION

CHECK LIST

BEAM OR
LOWER CHORD MEMBER
TUBE
FILLET FIELD WELD

LOWER CHORD MEMBER
OR
BEAM

COLUMN

PLAN VIEW

SLIDE4

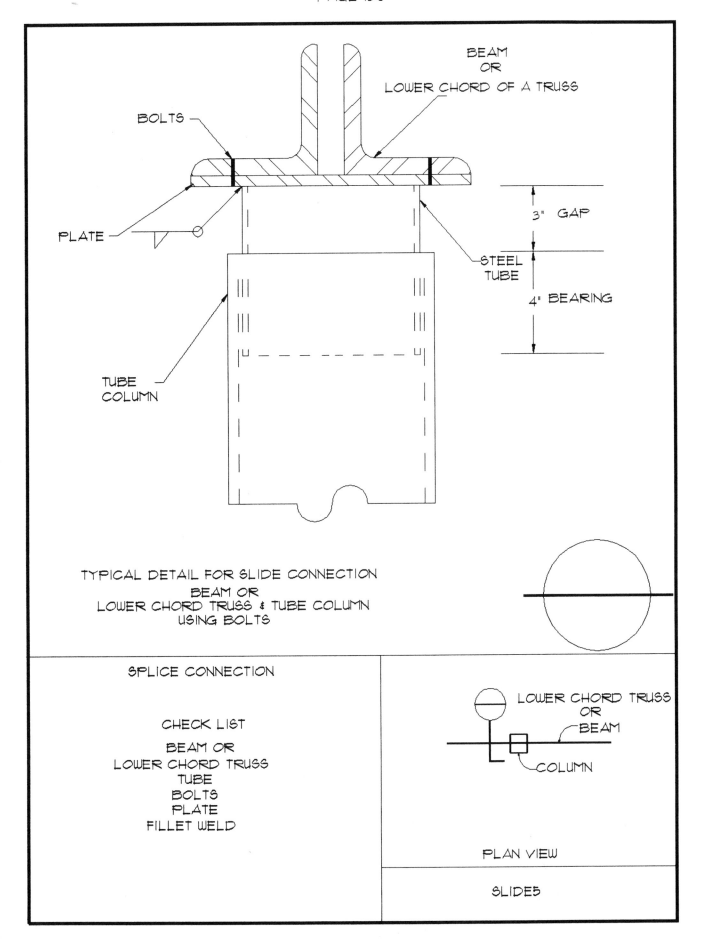

BEAM
OR
LOWER CHORD OF A TRUSS

BOLTS

PLATE

3" GAP

STEEL
TUBE

4" BEARING

TUBE
COLUMN

TYPICAL DETAIL FOR SLIDE CONNECTION
BEAM OR
LOWER CHORD TRUSS & TUBE COLUMN
USING BOLTS

SPLICE CONNECTION

CHECK LIST

BEAM OR
LOWER CHORD TRUSS
TUBE
BOLTS
PLATE
FILLET WELD

LOWER CHORD TRUSS
OR
BEAM

COLUMN

PLAN VIEW

SLIDE5

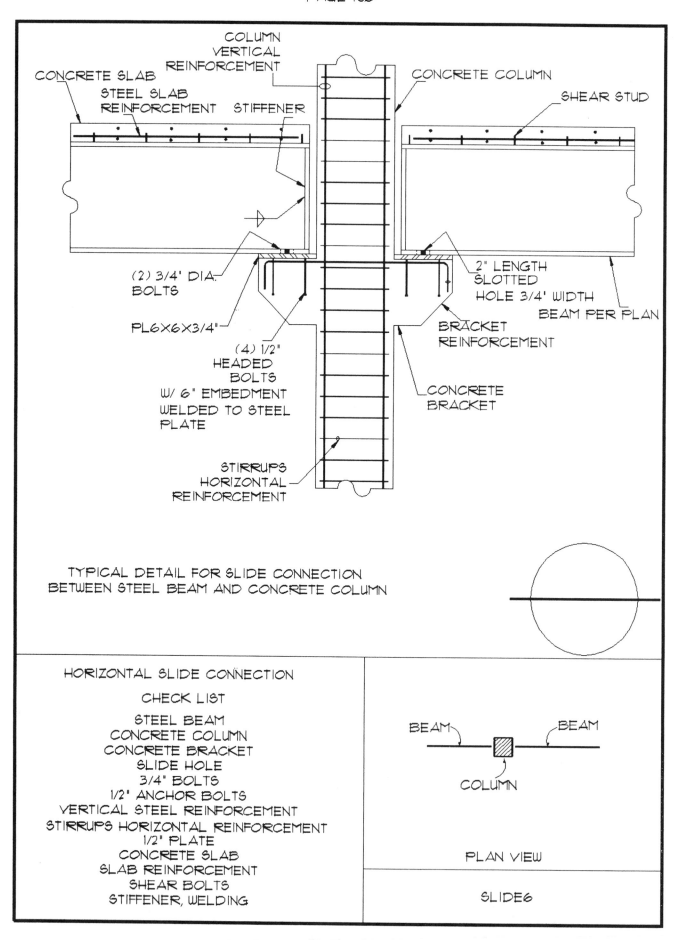

CONCRETE SLAB

STEEL SLAB
REINFORCEMENT

COLUMN
VERTICAL
REINFORCEMENT

STIFFENER

CONCRETE COLUMN

SHEAR STUD

(2) 3/4' DIA.
BOLTS

PL6X6X3/4"

(4) 1/2"
HEADED
BOLTS
W/ 6" EMBEDMENT
WELDED TO STEEL
PLATE

2" LENGTH
SLOTTED
HOLE 3/4' WIDTH

BEAM PER PLAN

BRACKET
REINFORCEMENT

CONCRETE
BRACKET

STIRRUPS
HORIZONTAL
REINFORCEMENT

TYPICAL DETAIL FOR SLIDE CONNECTION
BETWEEN STEEL BEAM AND CONCRETE COLUMN

HORIZONTAL SLIDE CONNECTION

CHECK LIST

STEEL BEAM
CONCRETE COLUMN
CONCRETE BRACKET
SLIDE HOLE
3/4" BOLTS
1/2" ANCHOR BOLTS
VERTICAL STEEL REINFORCEMENT
STIRRUPS HORIZONTAL REINFORCEMENT
1/2" PLATE
CONCRETE SLAB
SLAB REINFORCEMENT
SHEAR BOLTS
STIFFENER, WELDING

BEAM

BEAM

COLUMN

PLAN VIEW

SLIDE6

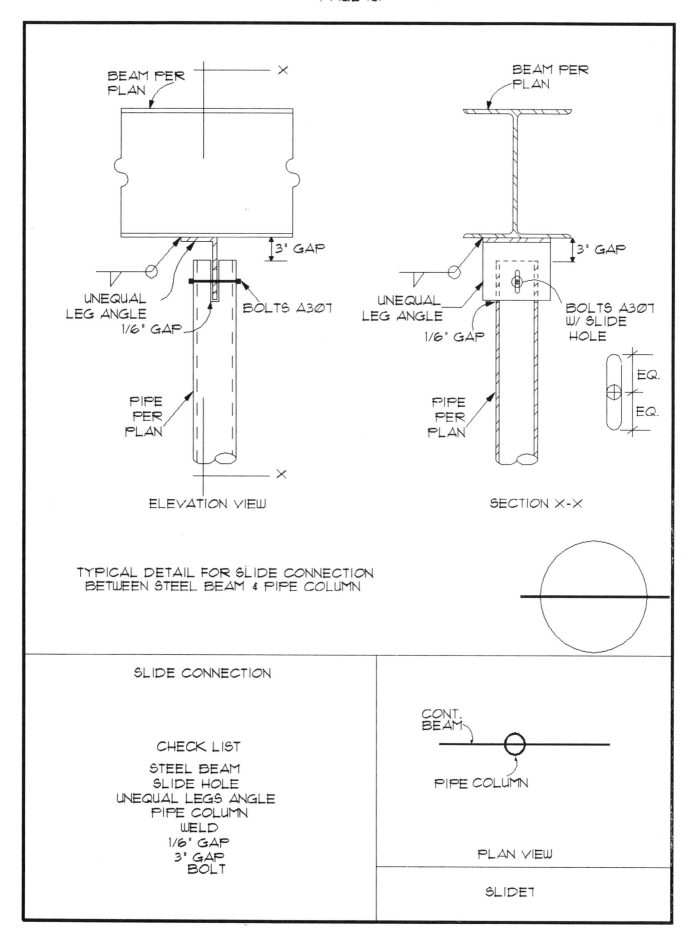

BEAM PER PLAN

X

3" GAP

UNEQUAL LEG ANGLE

1/6" GAP

BOLTS A307

PIPE PER PLAN

X

ELEVATION VIEW

BEAM PER PLAN

3" GAP

UNEQUAL LEG ANGLE

1/6" GAP

BOLTS A307 W/ SLIDE HOLE

PIPE PER PLAN

EQ.

EQ.

SECTION X-X

TYPICAL DETAIL FOR SLIDE CONNECTION
BETWEEN STEEL BEAM & PIPE COLUMN

SLIDE CONNECTION

CHECK LIST

STEEL BEAM
SLIDE HOLE
UNEQUAL LEGS ANGLE
PIPE COLUMN
WELD
1/6" GAP
3" GAP
BOLT

CONT. BEAM

PIPE COLUMN

PLAN VIEW

SLIDE7

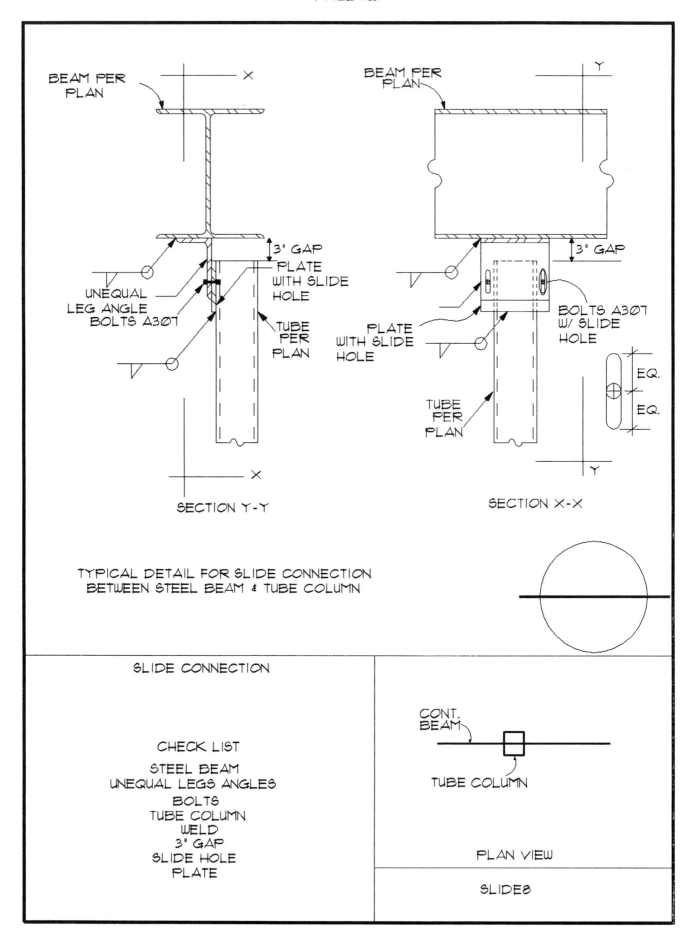

BEAM PER PLAN

3" GAP

PLATE WITH SLIDE HOLE

UNEQUAL LEG ANGLE BOLTS A307

TUBE PER PLAN

SECTION Y-Y

BEAM PER PLAN

3" GAP

PLATE WITH SLIDE HOLE

BOLTS A307 W/ SLIDE HOLE

TUBE PER PLAN

EQ.

EQ.

SECTION X-X

TYPICAL DETAIL FOR SLIDE CONNECTION
BETWEEN STEEL BEAM & TUBE COLUMN

SLIDE CONNECTION

CHECK LIST

STEEL BEAM
UNEQUAL LEGS ANGLES
BOLTS
TUBE COLUMN
WELD
3" GAP
SLIDE HOLE
PLATE

CONT. BEAM

TUBE COLUMN

PLAN VIEW

SLIDES

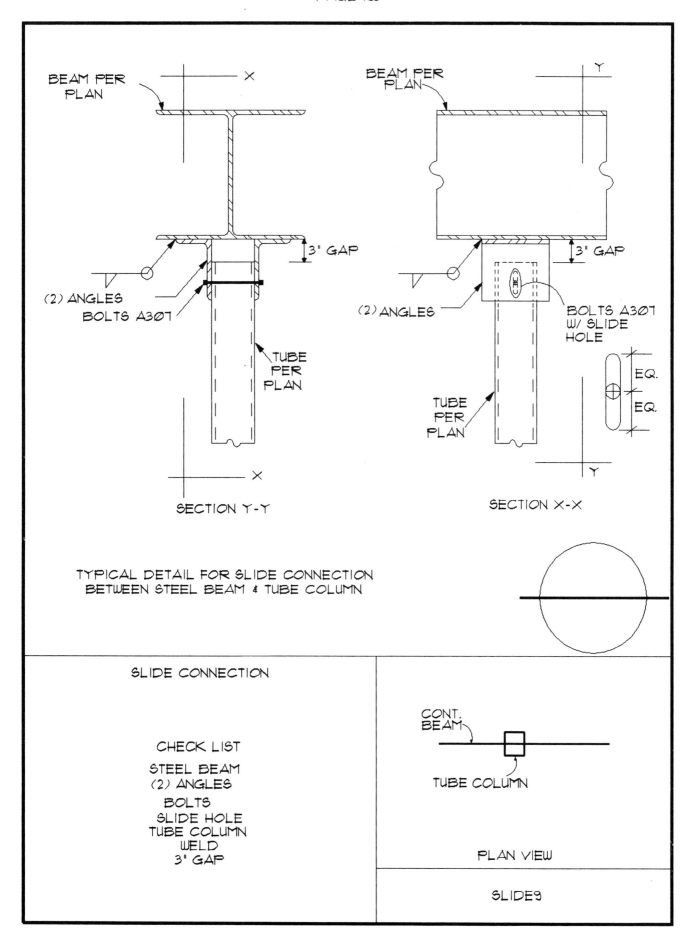

BEAM PER PLAN

X

(2) ANGLES

BOLTS A307

3" GAP

TUBE PER PLAN

SECTION Y-Y

BEAM PER PLAN

Y

(2) ANGLES

3" GAP

BOLTS A307 W/ SLIDE HOLE

TUBE PER PLAN

EQ.

EQ.

Y

SECTION X-X

TYPICAL DETAIL FOR SLIDE CONNECTION BETWEEN STEEL BEAM & TUBE COLUMN

SLIDE CONNECTION

CHECK LIST

STEEL BEAM
(2) ANGLES
BOLTS
SLIDE HOLE
TUBE COLUMN
WELD
3" GAP

CONT. BEAM

TUBE COLUMN

PLAN VIEW

SLIDE9

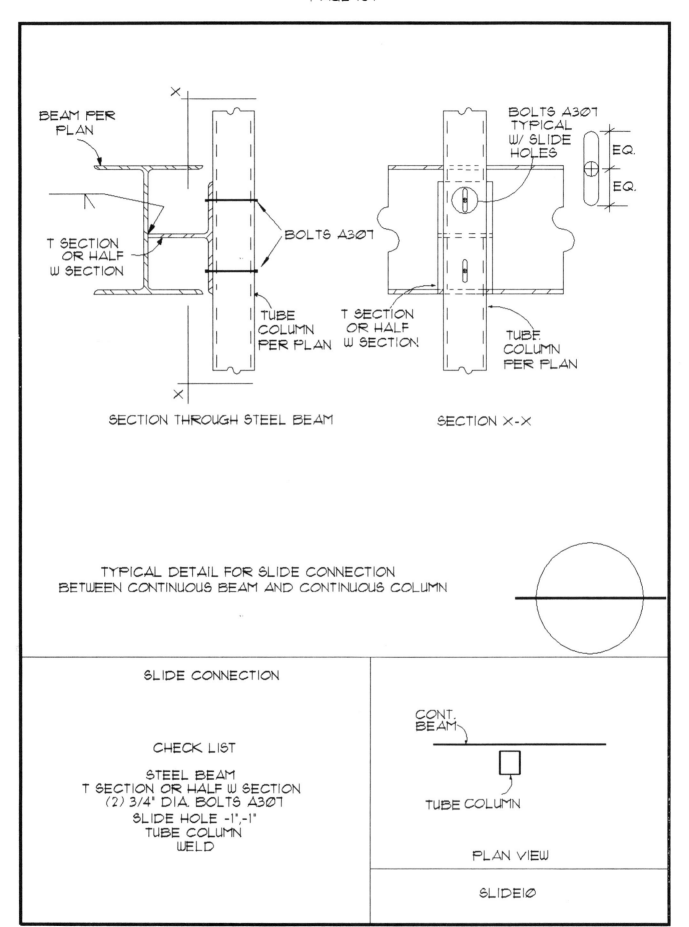

BEAM PER PLAN

T SECTION OR HALF W SECTION

BOLTS A307

TUBE COLUMN PER PLAN

SECTION THROUGH STEEL BEAM

BOLTS A307 TYPICAL W/ SLIDE HOLES

EQ.

EQ.

T SECTION OR HALF W SECTION

TUBE COLUMN PER PLAN

SECTION X-X

TYPICAL DETAIL FOR SLIDE CONNECTION
BETWEEN CONTINUOUS BEAM AND CONTINUOUS COLUMN

SLIDE CONNECTION

CHECK LIST

STEEL BEAM
T SECTION OR HALF W SECTION
(2) 3/4" DIA. BOLTS A307
SLIDE HOLE -1",-1"
TUBE COLUMN
WELD

CONT. BEAM

TUBE COLUMN

PLAN VIEW

SLIDE10

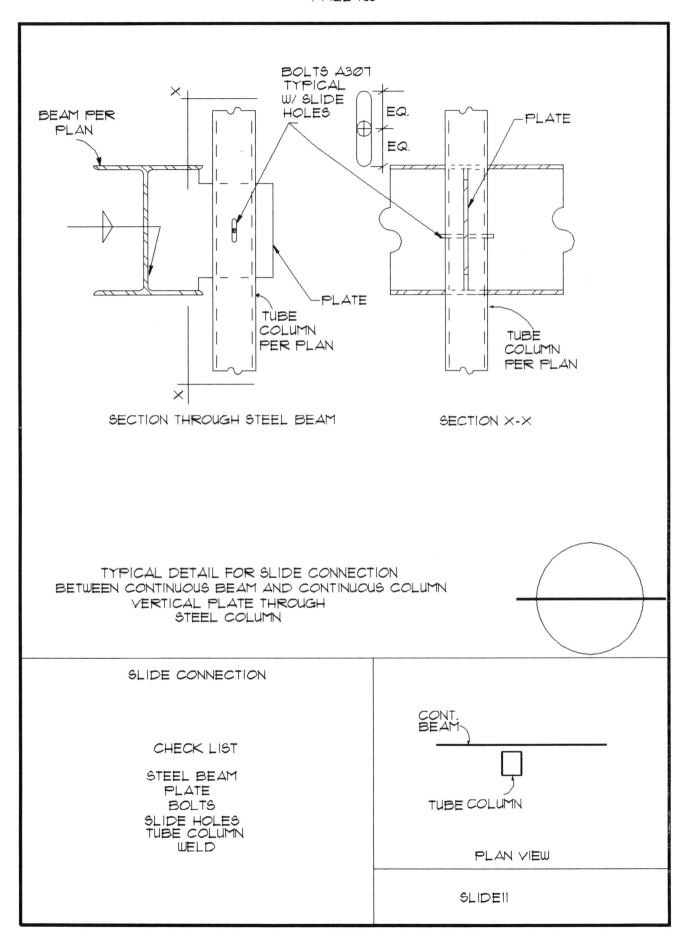

BEAM PER
PLAN

BOLTS A307
TYPICAL
W/ SLIDE
HOLES

EQ.

EQ.

PLATE

PLATE

TUBE
COLUMN
PER PLAN

TUBE
COLUMN
PER PLAN

SECTION THROUGH STEEL BEAM

SECTION X-X

TYPICAL DETAIL FOR SLIDE CONNECTION
BETWEEN CONTINUOUS BEAM AND CONTINUOUS COLUMN
VERTICAL PLATE THROUGH
STEEL COLUMN

SLIDE CONNECTION

CHECK LIST

STEEL BEAM
PLATE
BOLTS
SLIDE HOLES
TUBE COLUMN
WELD

CONT.
BEAM

TUBE COLUMN

PLAN VIEW

SLIDE11

CHAPTER 14

STEEL MEMBER SPLICE

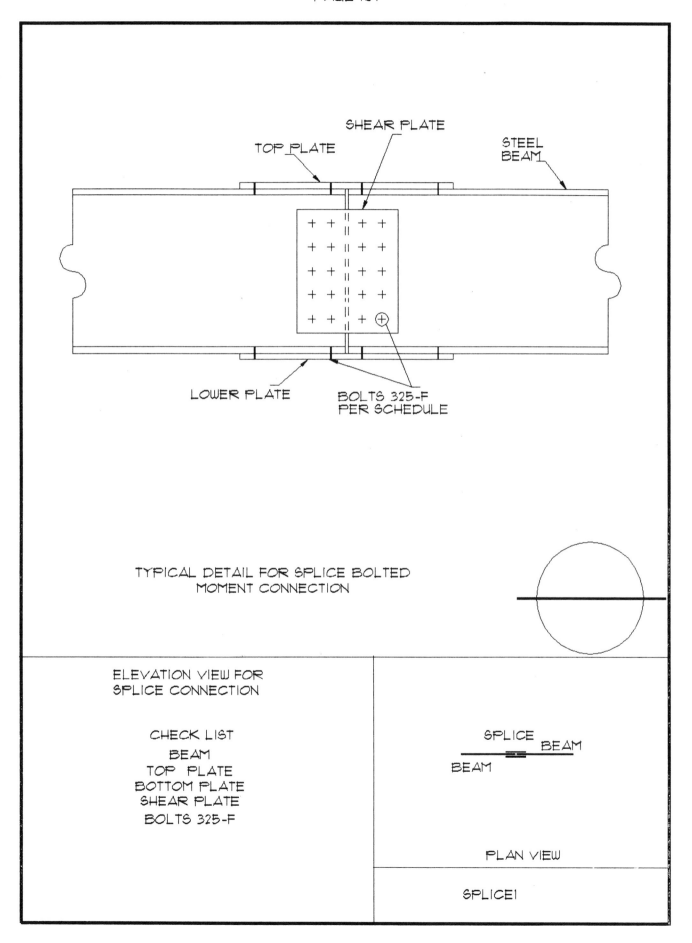

TOP PLATE

SHEAR PLATE

STEEL
BEAM

+ + + +
+ + + +
+ + + +
+ + + +
+ + + ⊕

LOWER PLATE

BOLTS 325-F
PER SCHEDULE

TYPICAL DETAIL FOR SPLICE BOLTED
MOMENT CONNECTION

ELEVATION VIEW FOR
SPLICE CONNECTION

CHECK LIST
BEAM
TOP PLATE
BOTTOM PLATE
SHEAR PLATE
BOLTS 325-F

SPLICE
BEAM
BEAM

PLAN VIEW

SPLICE1

PAGE 198

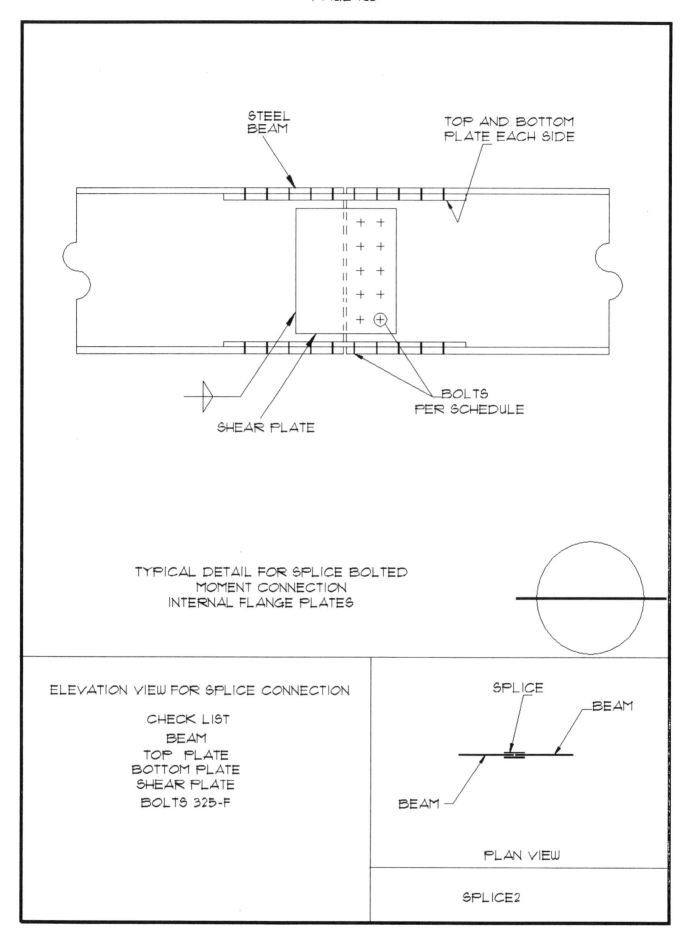

STEEL
BEAM

TOP AND BOTTOM
PLATE EACH SIDE

BOLTS
PER SCHEDULE

SHEAR PLATE

TYPICAL DETAIL FOR SPLICE BOLTED
MOMENT CONNECTION
INTERNAL FLANGE PLATES

ELEVATION VIEW FOR SPLICE CONNECTION

CHECK LIST
BEAM
TOP PLATE
BOTTOM PLATE
SHEAR PLATE
BOLTS 325-F

SPLICE

BEAM

BEAM

PLAN VIEW

SPLICE2

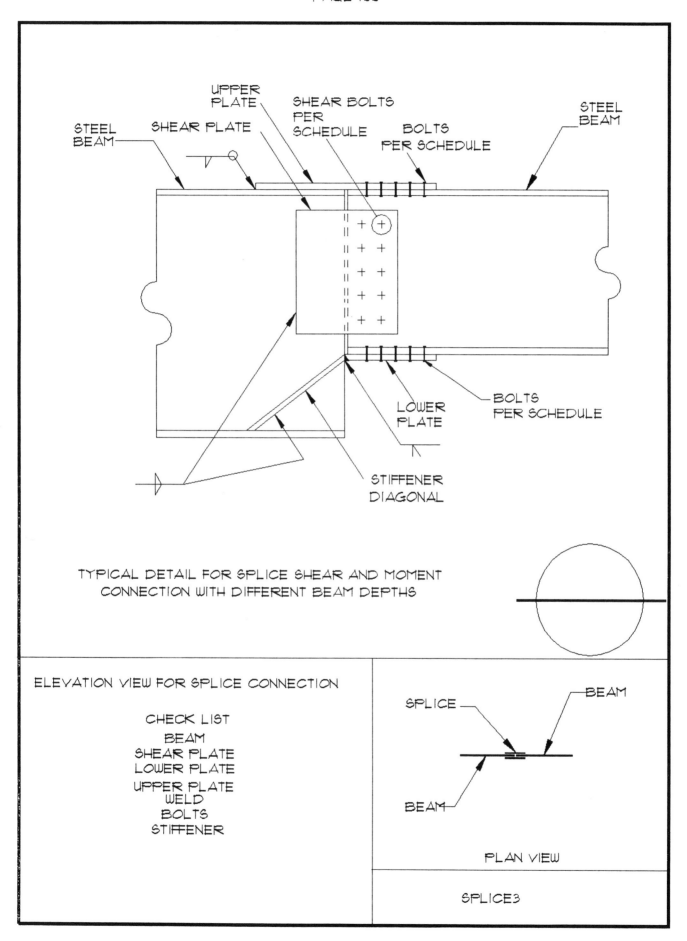

STEEL BEAM

SHEAR PLATE

UPPER PLATE

SHEAR BOLTS PER SCHEDULE

BOLTS PER SCHEDULE

STEEL BEAM

BOLTS PER SCHEDULE

LOWER PLATE

STIFFENER DIAGONAL

TYPICAL DETAIL FOR SPLICE SHEAR AND MOMENT
CONNECTION WITH DIFFERENT BEAM DEPTHS

ELEVATION VIEW FOR SPLICE CONNECTION

CHECK LIST
BEAM
SHEAR PLATE
LOWER PLATE
UPPER PLATE
WELD
BOLTS
STIFFENER

SPLICE

BEAM

BEAM

PLAN VIEW

SPLICE3

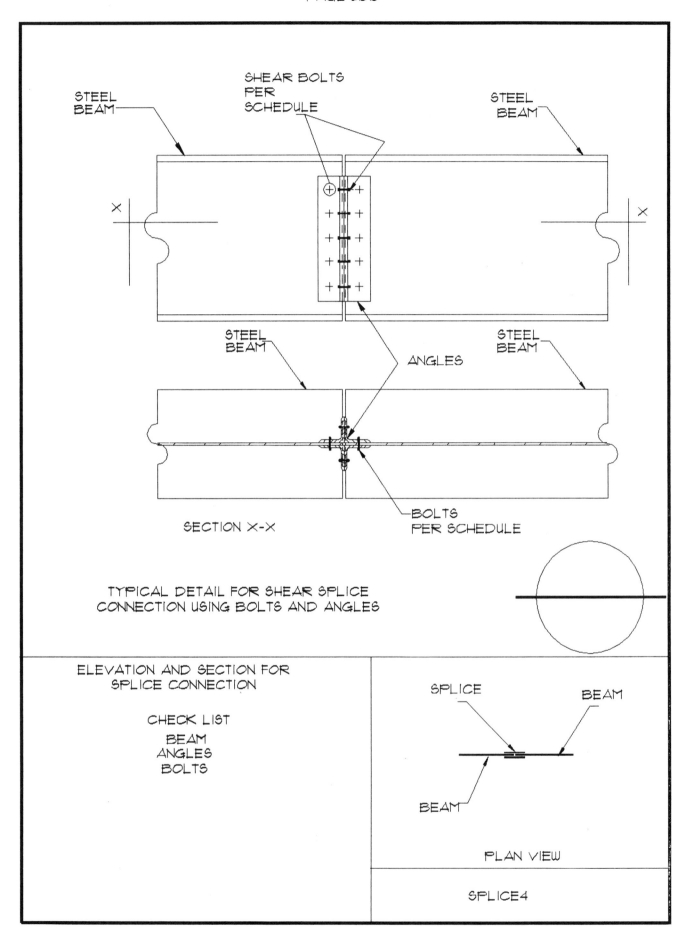

STEEL BEAM

SHEAR BOLTS PER SCHEDULE

STEEL BEAM

X

X

STEEL BEAM

ANGLES

STEEL BEAM

SECTION X-X

BOLTS PER SCHEDULE

TYPICAL DETAIL FOR SHEAR SPLICE
CONNECTION USING BOLTS AND ANGLES

ELEVATION AND SECTION FOR
SPLICE CONNECTION

CHECK LIST
BEAM
ANGLES
BOLTS

SPLICE

BEAM

BEAM

PLAN VIEW

SPLICE4

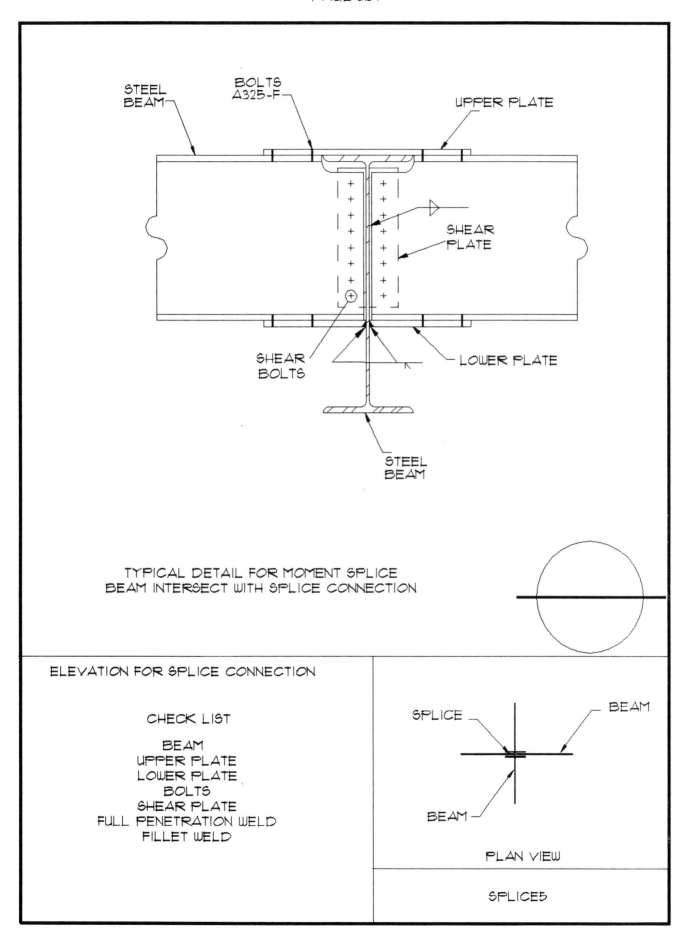

STEEL BEAM

BOLTS A325-F

UPPER PLATE

SHEAR PLATE

SHEAR BOLTS

LOWER PLATE

STEEL BEAM

TYPICAL DETAIL FOR MOMENT SPLICE
BEAM INTERSECT WITH SPLICE CONNECTION

ELEVATION FOR SPLICE CONNECTION

CHECK LIST

BEAM
UPPER PLATE
LOWER PLATE
BOLTS
SHEAR PLATE
FULL PENETRATION WELD
FILLET WELD

SPLICE

BEAM

BEAM

PLAN VIEW

SPLICE5

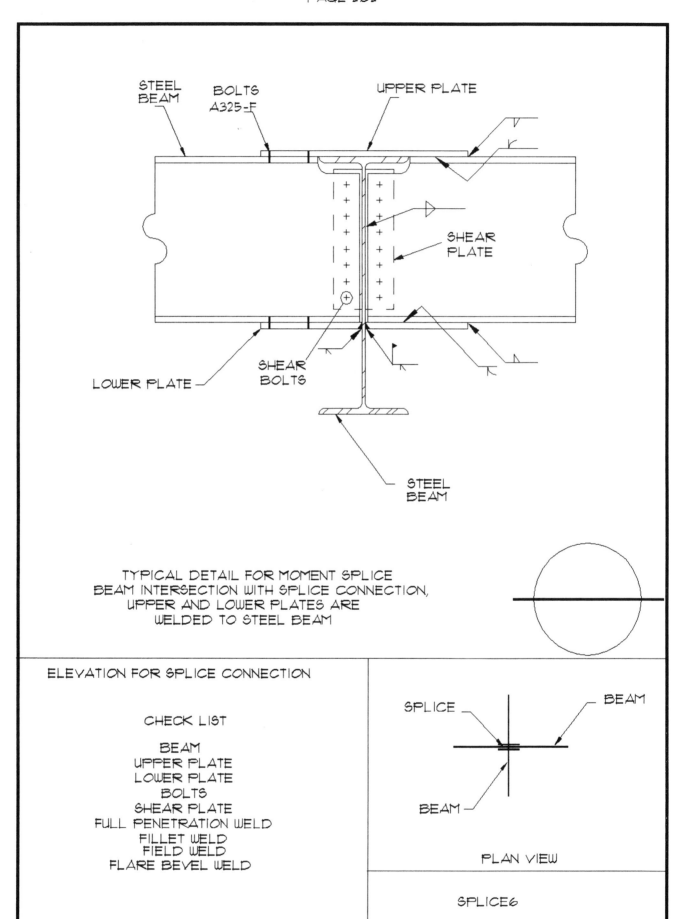

STEEL BEAM

BOLTS A325-F

UPPER PLATE

SHEAR PLATE

LOWER PLATE

SHEAR BOLTS

STEEL BEAM

TYPICAL DETAIL FOR MOMENT SPLICE
BEAM INTERSECTION WITH SPLICE CONNECTION,
UPPER AND LOWER PLATES ARE
WELDED TO STEEL BEAM

ELEVATION FOR SPLICE CONNECTION

CHECK LIST

BEAM
UPPER PLATE
LOWER PLATE
BOLTS
SHEAR PLATE
FULL PENETRATION WELD
FILLET WELD
FIELD WELD
FLARE BEVEL WELD

SPLICE

BEAM

BEAM

PLAN VIEW

SPLICE6

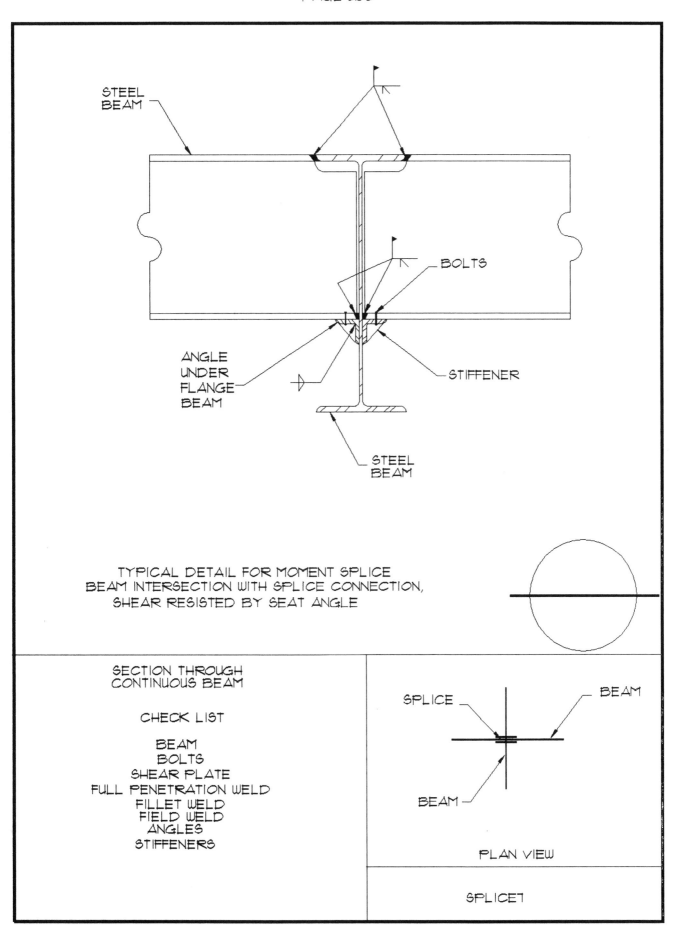

STEEL
BEAM

BOLTS

ANGLE
UNDER
FLANGE
BEAM

STIFFENER

STEEL
BEAM

TYPICAL DETAIL FOR MOMENT SPLICE
BEAM INTERSECTION WITH SPLICE CONNECTION,
SHEAR RESISTED BY SEAT ANGLE

SECTION THROUGH
CONTINUOUS BEAM

CHECK LIST

BEAM
BOLTS
SHEAR PLATE
FULL PENETRATION WELD
FILLET WELD
FIELD WELD
ANGLES
STIFFENERS

SPLICE

BEAM

BEAM

PLAN VIEW

SPLICET

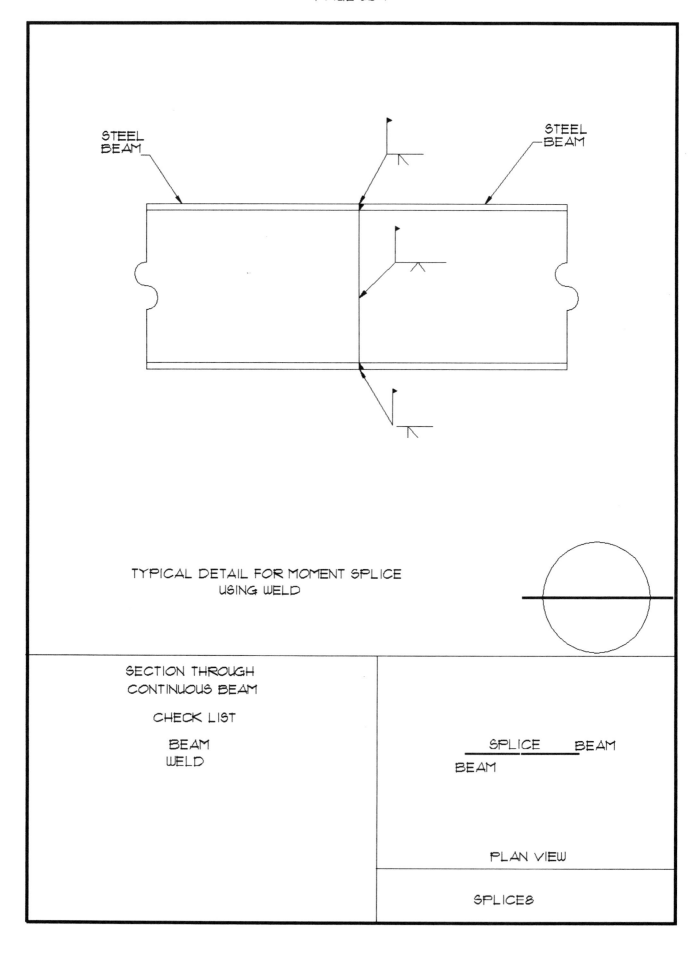

STEEL
BEAM

STEEL
BEAM

TYPICAL DETAIL FOR MOMENT SPLICE
USING WELD

SECTION THROUGH
CONTINUOUS BEAM

CHECK LIST

BEAM
WELD

SPLICE   BEAM

BEAM

PLAN VIEW

SPLICES

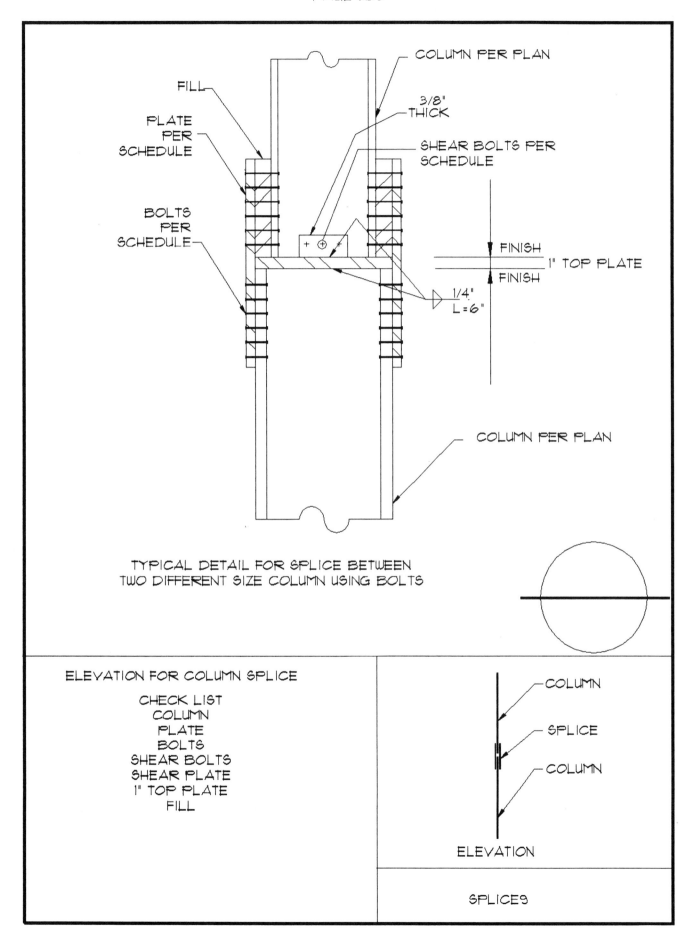

COLUMN PER PLAN

FILL

3/8" THICK

PLATE PER SCHEDULE

SHEAR BOLTS PER SCHEDULE

BOLTS PER SCHEDULE

FINISH

1" TOP PLATE

FINISH

1/4" L=6"

COLUMN PER PLAN

TYPICAL DETAIL FOR SPLICE BETWEEN
TWO DIFFERENT SIZE COLUMN USING BOLTS

ELEVATION FOR COLUMN SPLICE

CHECK LIST
COLUMN
PLATE
BOLTS
SHEAR BOLTS
SHEAR PLATE
1" TOP PLATE
FILL

COLUMN

SPLICE

COLUMN

ELEVATION

SPLICES

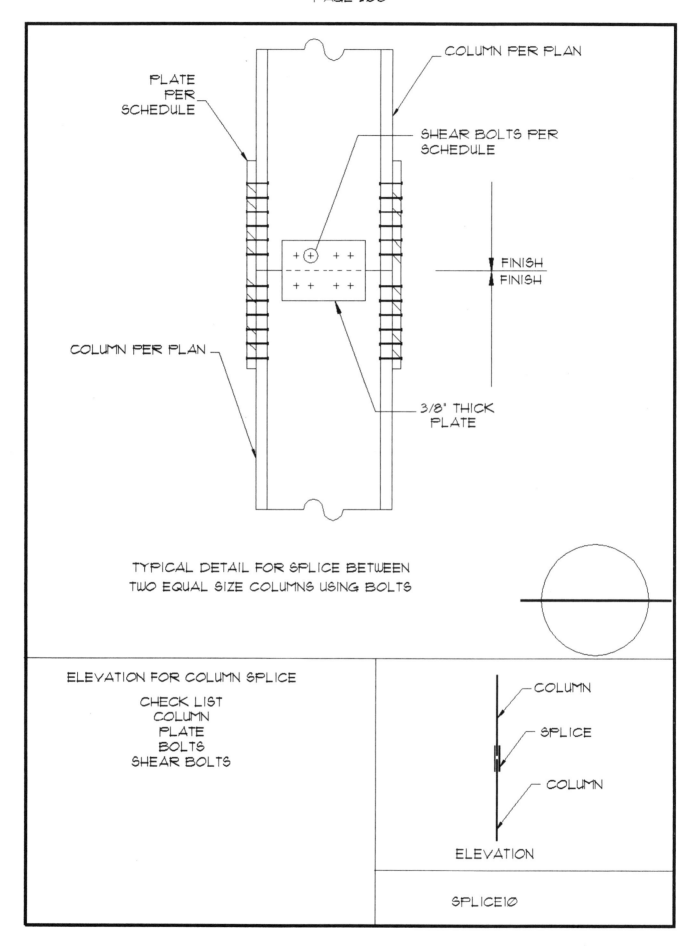

COLUMN PER PLAN

PLATE
PER
SCHEDULE

SHEAR BOLTS PER
SCHEDULE

FINISH
FINISH

COLUMN PER PLAN

3/8" THICK
PLATE

TYPICAL DETAIL FOR SPLICE BETWEEN
TWO EQUAL SIZE COLUMNS USING BOLTS

ELEVATION FOR COLUMN SPLICE

CHECK LIST
COLUMN
PLATE
BOLTS
SHEAR BOLTS

COLUMN

SPLICE

COLUMN

ELEVATION

SPLICE10

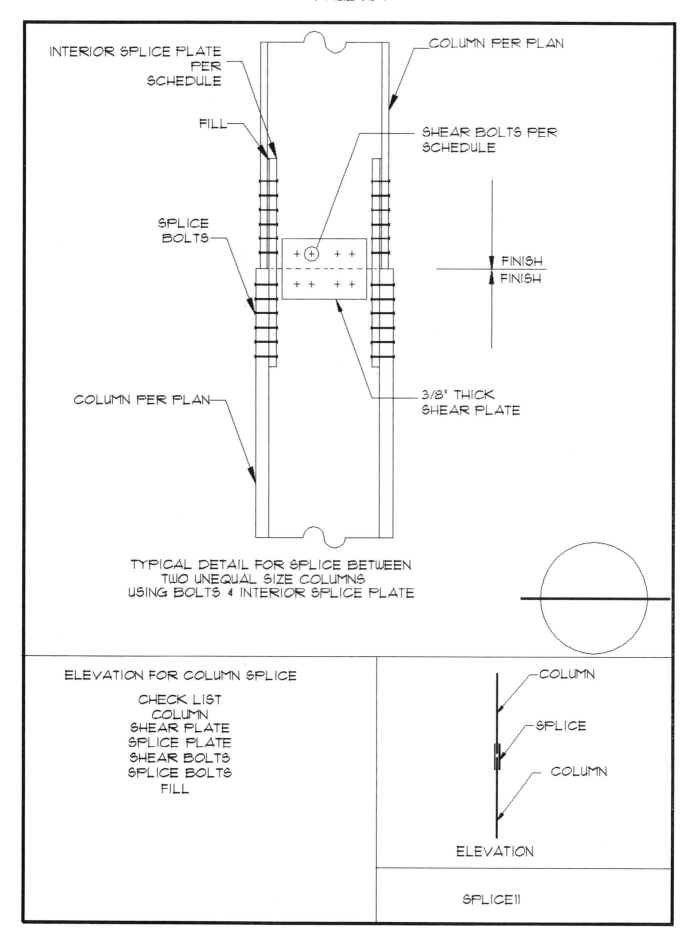

INTERIOR SPLICE PLATE PER SCHEDULE

COLUMN PER PLAN

FILL

SHEAR BOLTS PER SCHEDULE

SPLICE BOLTS

FINISH
FINISH

COLUMN PER PLAN

3/8" THICK SHEAR PLATE

TYPICAL DETAIL FOR SPLICE BETWEEN
TWO UNEQUAL SIZE COLUMNS
USING BOLTS & INTERIOR SPLICE PLATE

ELEVATION FOR COLUMN SPLICE

CHECK LIST
COLUMN
SHEAR PLATE
SPLICE PLATE
SHEAR BOLTS
SPLICE BOLTS
FILL

COLUMN

SPLICE

COLUMN

ELEVATION

SPLICE11

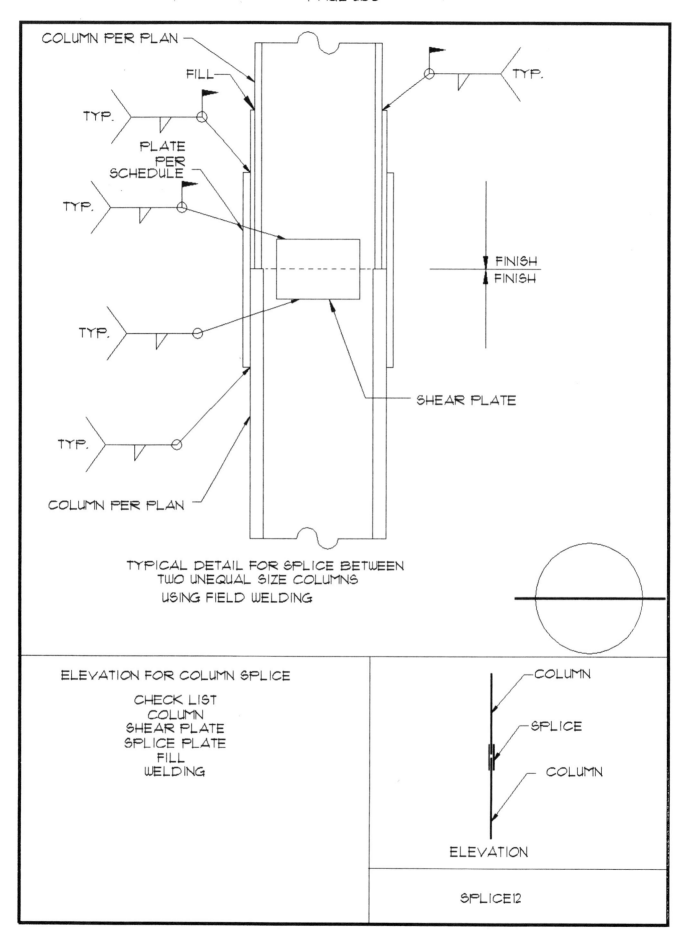

COLUMN PER PLAN

FILL

TYP.

TYP.

PLATE
PER
SCHEDULE

TYP.

FINISH
FINISH

TYP.

SHEAR PLATE

TYP.

COLUMN PER PLAN

TYPICAL DETAIL FOR SPLICE BETWEEN
TWO UNEQUAL SIZE COLUMNS
USING FIELD WELDING

ELEVATION FOR COLUMN SPLICE

CHECK LIST
COLUMN
SHEAR PLATE
SPLICE PLATE
FILL
WELDING

COLUMN

SPLICE

COLUMN

ELEVATION

SPLICE12

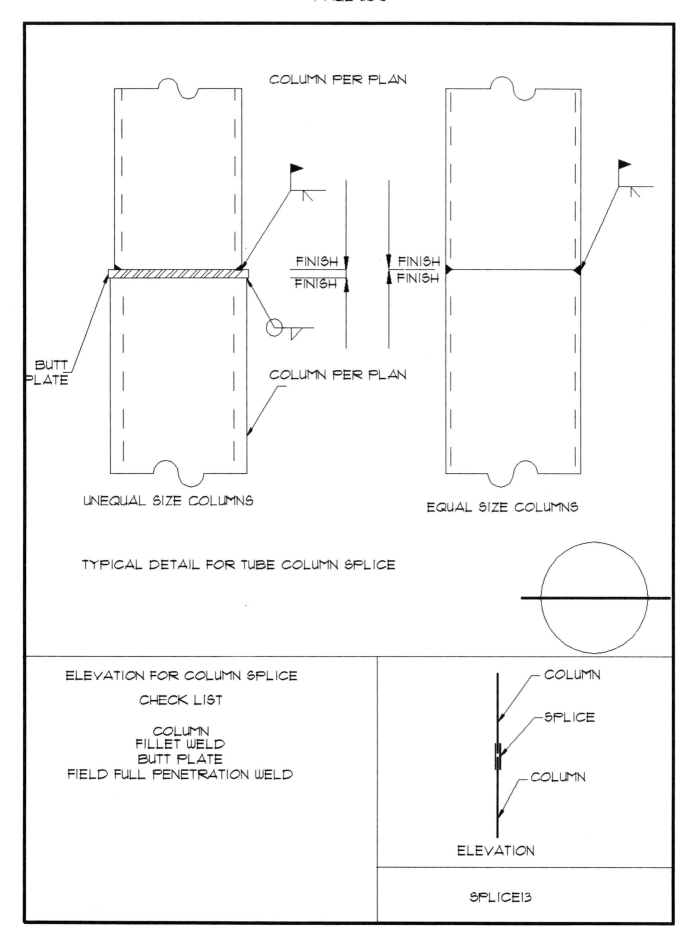

COLUMN PER PLAN

FINISH
FINISH

FINISH
FINISH

BUTT
PLATE

COLUMN PER PLAN

UNEQUAL SIZE COLUMNS

EQUAL SIZE COLUMNS

TYPICAL DETAIL FOR TUBE COLUMN SPLICE

ELEVATION FOR COLUMN SPLICE

CHECK LIST

COLUMN
FILLET WELD
BUTT PLATE
FIELD FULL PENETRATION WELD

COLUMN

SPLICE

COLUMN

ELEVATION

SPLICE13

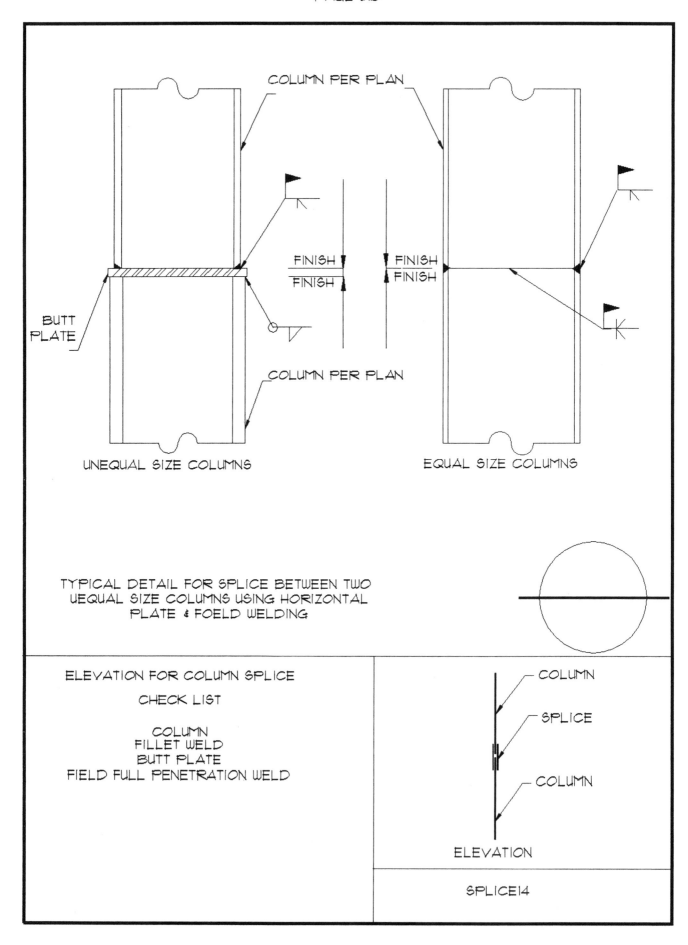

COLUMN PER PLAN

BUTT
PLATE

FINISH
FINISH

FINISH
FINISH

COLUMN PER PLAN

UNEQUAL SIZE COLUMNS

EQUAL SIZE COLUMNS

TYPICAL DETAIL FOR SPLICE BETWEEN TWO
UEQUAL SIZE COLUMNS USING HORIZONTAL
PLATE & FOELD WELDING

ELEVATION FOR COLUMN SPLICE

CHECK LIST

COLUMN
FILLET WELD
BUTT PLATE
FIELD FULL PENETRATION WELD

COLUMN

SPLICE

COLUMN

ELEVATION

SPLICE14

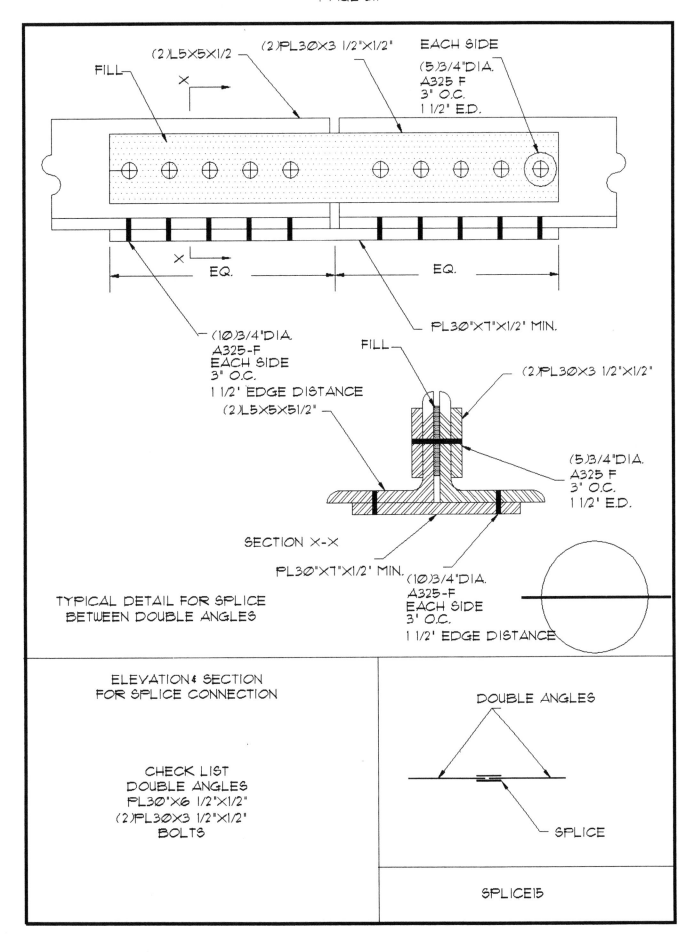

(2)L5X5X1/2

(2)PL30X3 1/2"X1/2"

EACH SIDE
(5)3/4"DIA.
A325 F
3" O.C.
1 1/2" E.D.

FILL

X

EQ.

X

EQ.

PL30"X7"X1/2" MIN.

(10)3/4"DIA.
A325-F
EACH SIDE
3" O.C.
1 1/2' EDGE DISTANCE

(2)L5X5X51/2"

FILL

(2)PL30X3 1/2"X1/2"

(5)3/4"DIA.
A325 F
3" O.C.
1 1/2' E.D.

SECTION X-X

PL30"X7"X1/2' MIN.

(10)3/4"DIA.
A325-F
EACH SIDE
3' O.C.
1 1/2' EDGE DISTANCE

TYPICAL DETAIL FOR SPLICE
BETWEEN DOUBLE ANGLES

ELEVATION & SECTION
FOR SPLICE CONNECTION

CHECK LIST
DOUBLE ANGLES
PL30"X6 1/2"X1/2"
(2)PL30X3 1/2"X1/2"
BOLTS

DOUBLE ANGLES

SPLICE

SPLICE15

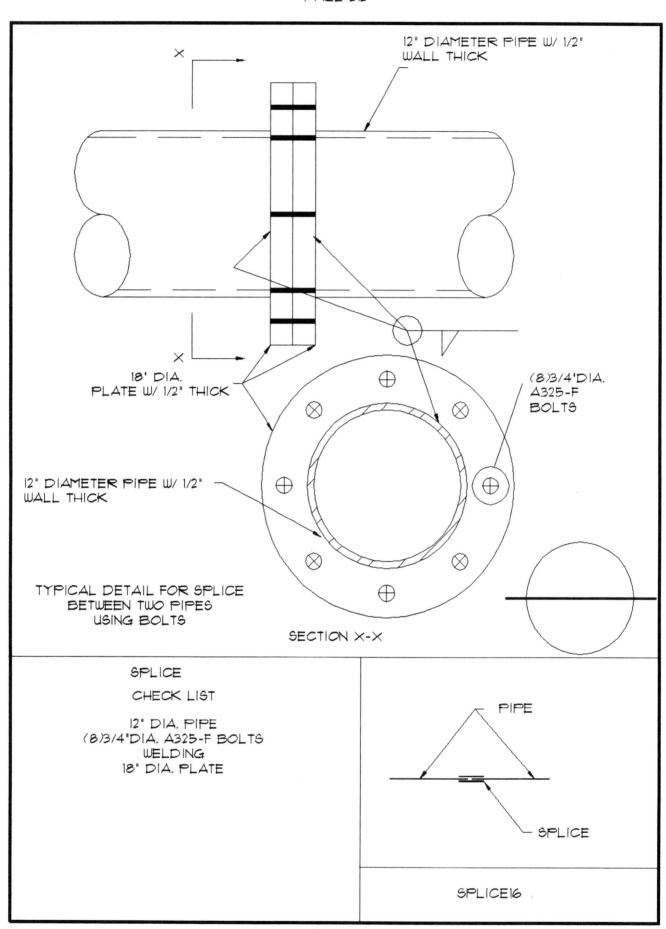

12" DIAMETER PIPE W/ 1/2" WALL THICK

18' DIA. PLATE W/ 1/2" THICK

(8)3/4'DIA. A325-F BOLTS

12" DIAMETER PIPE W/ 1/2" WALL THICK

TYPICAL DETAIL FOR SPLICE BETWEEN TWO PIPES USING BOLTS

SECTION X-X

SPLICE

CHECK LIST

12" DIA. PIPE
(8)3/4"DIA. A325-F BOLTS
WELDING
18" DIA. PLATE

PIPE

SPLICE

SPLICE16

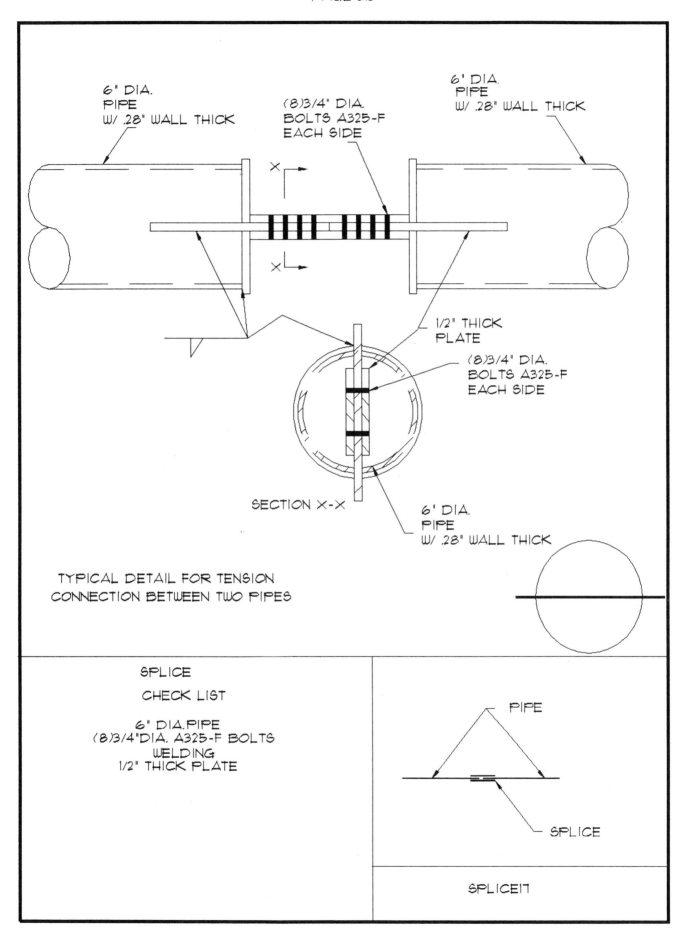

6" DIA.
PIPE
W/ .28" WALL THICK

(8)3/4" DIA.
BOLTS A325-F
EACH SIDE

6' DIA.
PIPE
W/ .28" WALL THICK

X

X

1/2" THICK
PLATE

(8)3/4" DIA.
BOLTS A325-F
EACH SIDE

SECTION X-X

6' DIA.
PIPE
W/ .28" WALL THICK

TYPICAL DETAIL FOR TENSION
CONNECTION BETWEEN TWO PIPES

SPLICE

CHECK LIST

6" DIA.PIPE
(8)3/4"DIA. A325-F BOLTS
WELDING
1/2" THICK PLATE

PIPE

SPLICE

SPLICE17

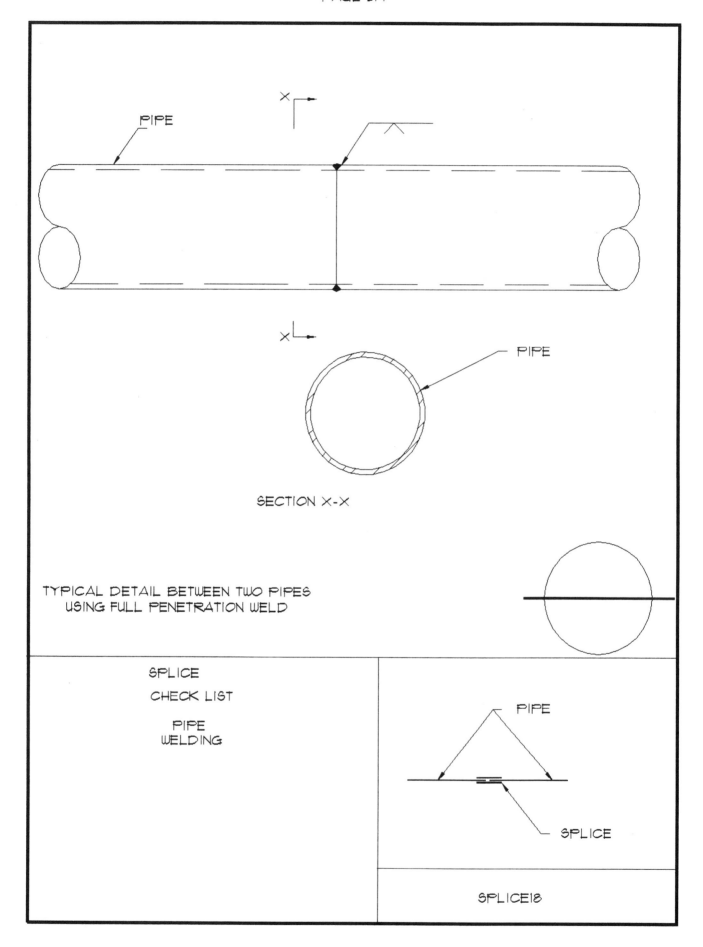

PIPE

X

PIPE

SECTION X-X

TYPICAL DETAIL BETWEEN TWO PIPES
USING FULL PENETRATION WELD

SPLICE

CHECK LIST

PIPE
WELDING

PIPE

SPLICE

SPLICE18

CHAPTER 15

STEEL STAIR DETAILS

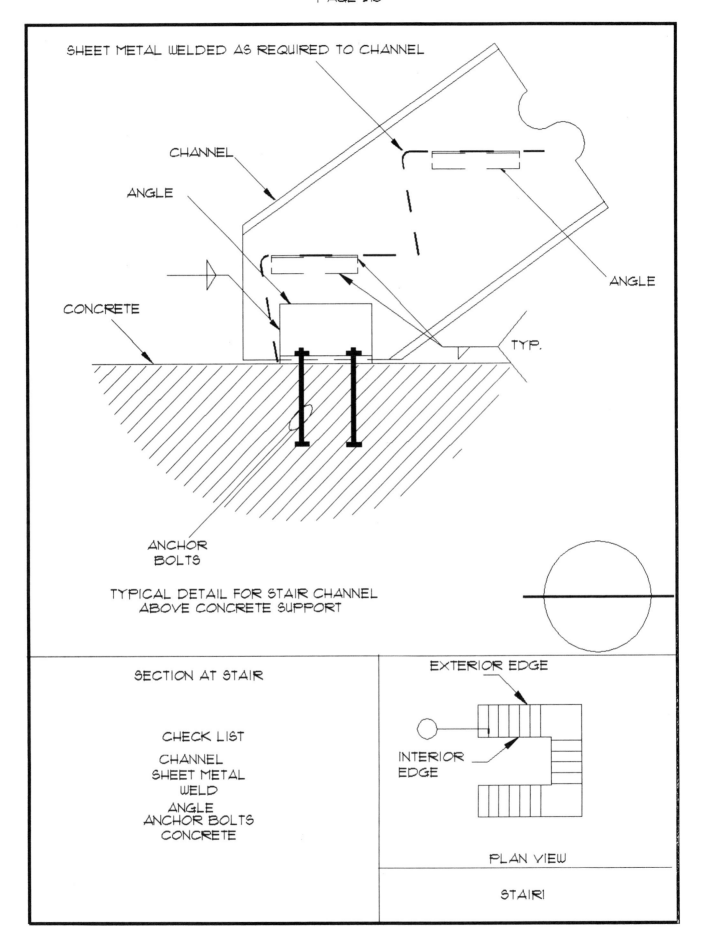

SHEET METAL WELDED AS REQUIRED TO CHANNEL

CHANNEL

ANGLE

ANGLE

CONCRETE

TYP.

ANCHOR
BOLTS

TYPICAL DETAIL FOR STAIR CHANNEL
ABOVE CONCRETE SUPPORT

SECTION AT STAIR

CHECK LIST

CHANNEL
SHEET METAL
WELD
ANGLE
ANCHOR BOLTS
CONCRETE

EXTERIOR EDGE

INTERIOR
EDGE

PLAN VIEW

STAIR1

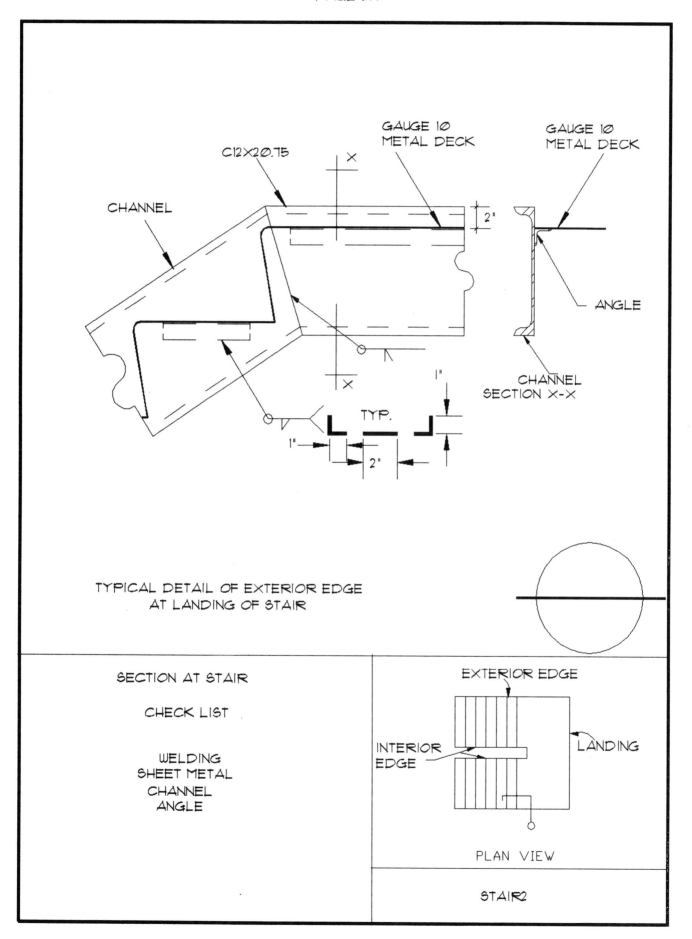

CHANNEL

C12X20.75

GAUGE 10
METAL DECK

GAUGE 10
METAL DECK

2"

ANGLE

CHANNEL
SECTION X-X

X

X

1"

TYP.

1"

2"

TYPICAL DETAIL OF EXTERIOR EDGE
AT LANDING OF STAIR

SECTION AT STAIR

CHECK LIST

WELDING
SHEET METAL
CHANNEL
ANGLE

EXTERIOR EDGE

INTERIOR
EDGE

LANDING

PLAN VIEW

STAIR2

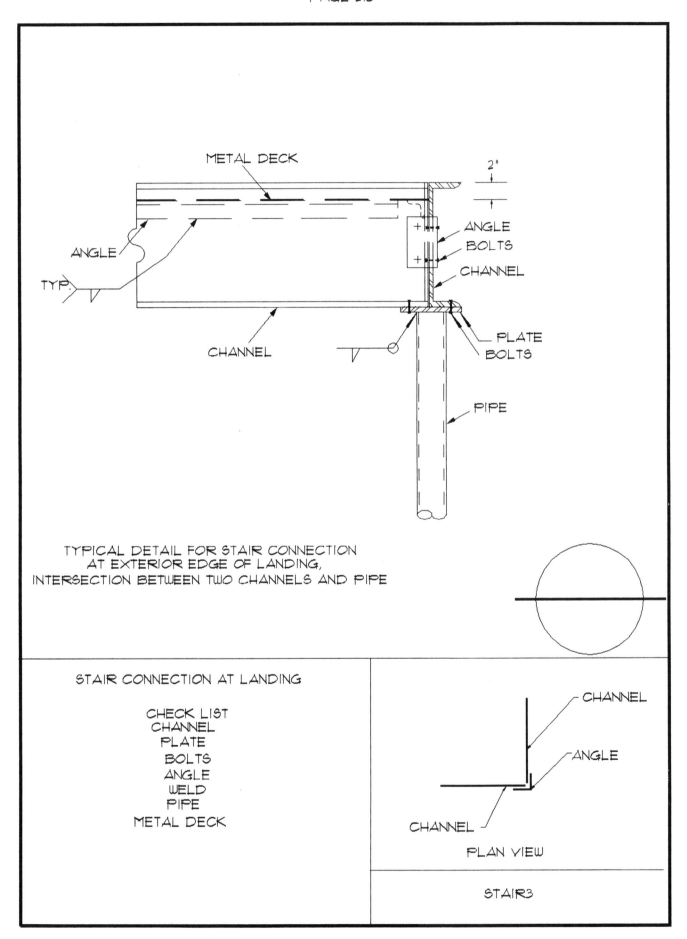

METAL DECK

2"

ANGLE

BOLTS

ANGLE

CHANNEL

TYP.

PLATE

BOLTS

CHANNEL

PIPE

TYPICAL DETAIL FOR STAIR CONNECTION
AT EXTERIOR EDGE OF LANDING,
INTERSECTION BETWEEN TWO CHANNELS AND PIPE

STAIR CONNECTION AT LANDING

CHECK LIST
CHANNEL
PLATE
BOLTS
ANGLE
WELD
PIPE
METAL DECK

CHANNEL

ANGLE

CHANNEL

PLAN VIEW

STAIR3

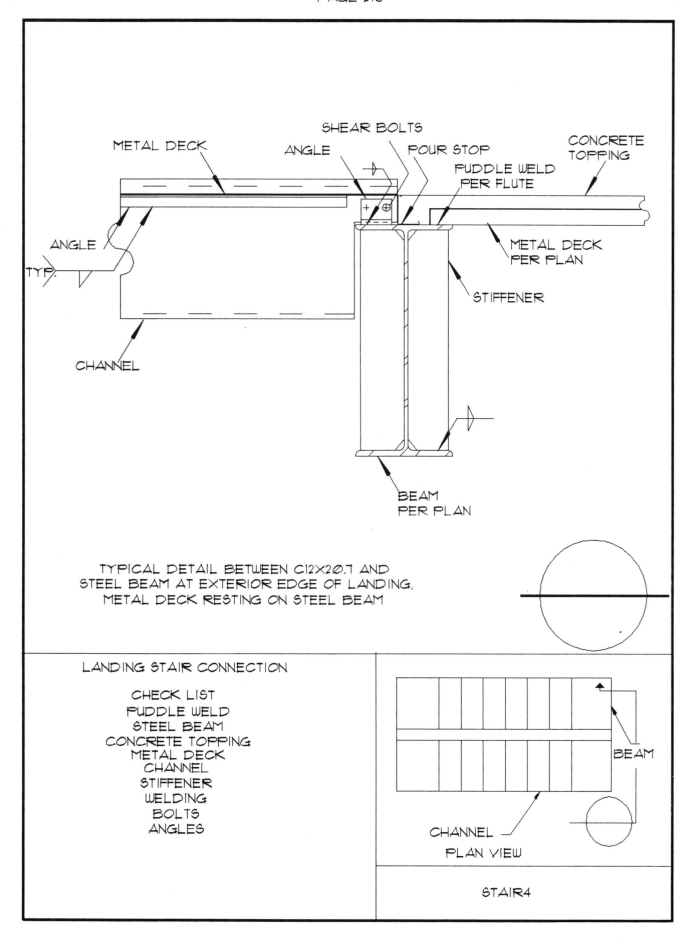

METAL DECK

ANGLE

SHEAR BOLTS

POUR STOP

CONCRETE TOPPING

PUDDLE WELD PER FLUTE

METAL DECK PER PLAN

STIFFENER

ANGLE

TYP.

CHANNEL

BEAM PER PLAN

TYPICAL DETAIL BETWEEN C12X20.7 AND STEEL BEAM AT EXTERIOR EDGE OF LANDING, METAL DECK RESTING ON STEEL BEAM

LANDING STAIR CONNECTION

CHECK LIST
PUDDLE WELD
STEEL BEAM
CONCRETE TOPPING
METAL DECK
CHANNEL
STIFFENER
WELDING
BOLTS
ANGLES

BEAM

CHANNEL

PLAN VIEW

STAIR4

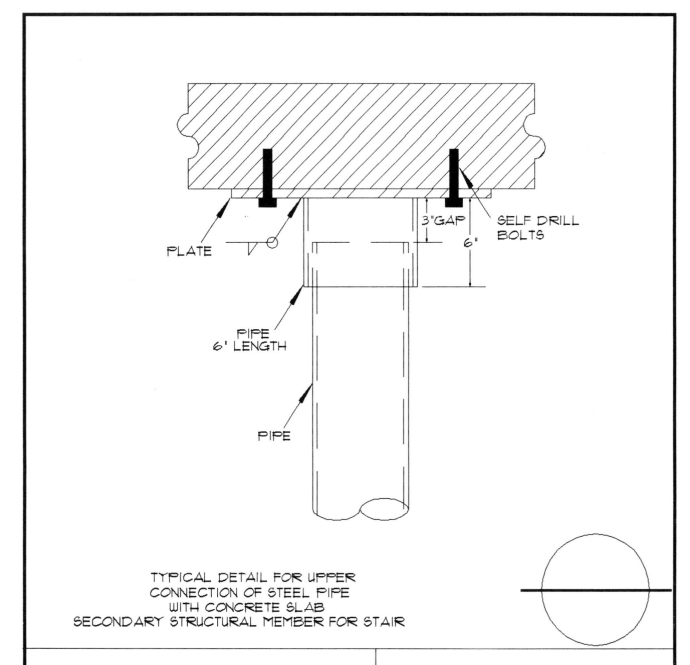

PLATE

3"GAP

SELF DRILL BOLTS

6"

PIPE 6' LENGTH

PIPE

TYPICAL DETAIL FOR UPPER
CONNECTION OF STEEL PIPE
WITH CONCRETE SLAB
SECONDARY STRUCTURAL MEMBER FOR STAIR

SECTION THROUGH CONCRETE SLAB

CHECK LIST
CONCRETE SLAB
STEEL REINFORCEMENT
SELF DRILL BOLTS
PIPE
3" GAP
WELDING

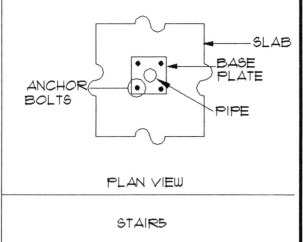

SLAB

BASE PLATE

ANCHOR BOLTS

PIPE

PLAN VIEW

STAIRS

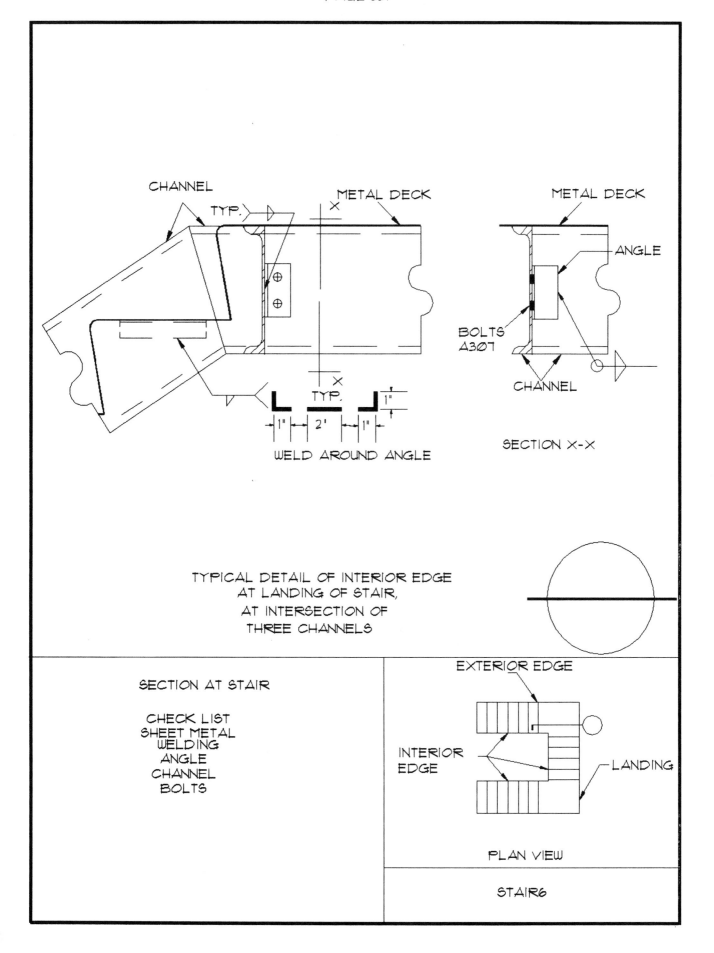

CHANNEL

TYP.

METAL DECK

WELD AROUND ANGLE

1"

1" 2' 1"

METAL DECK

ANGLE

BOLTS
A307

CHANNEL

SECTION X-X

TYPICAL DETAIL OF INTERIOR EDGE
AT LANDING OF STAIR,
AT INTERSECTION OF
THREE CHANNELS

SECTION AT STAIR

CHECK LIST
SHEET METAL
WELDING
ANGLE
CHANNEL
BOLTS

EXTERIOR EDGE

INTERIOR
EDGE

LANDING

PLAN VIEW

STAIR6

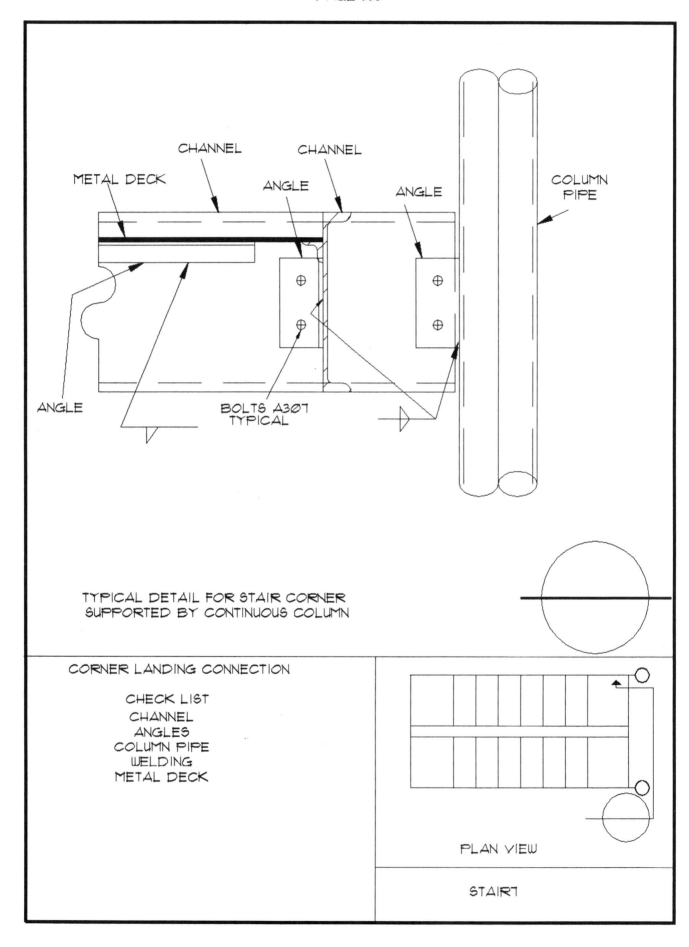

CHANNEL

CHANNEL

METAL DECK

ANGLE

ANGLE

COLUMN PIPE

ANGLE

BOLTS A307 TYPICAL

TYPICAL DETAIL FOR STAIR CORNER
SUPPORTED BY CONTINUOUS COLUMN

CORNER LANDING CONNECTION

CHECK LIST
CHANNEL
ANGLES
COLUMN PIPE
WELDING
METAL DECK

PLAN VIEW

STAIRT

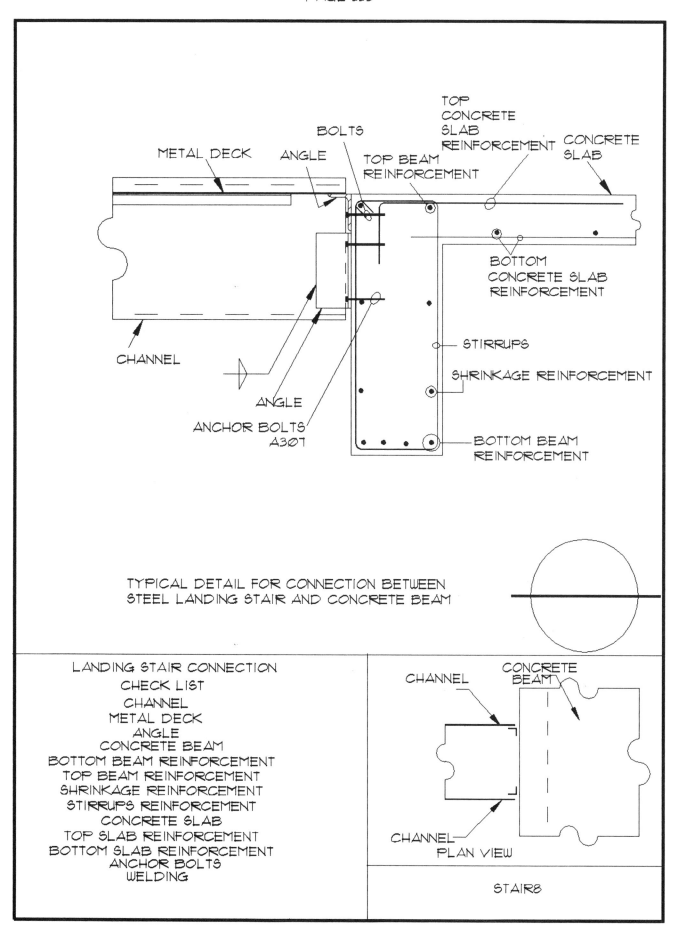

METAL DECK

ANGLE

BOLTS

TOP BEAM REINFORCEMENT

TOP CONCRETE SLAB REINFORCEMENT

CONCRETE SLAB

BOTTOM CONCRETE SLAB REINFORCEMENT

STIRRUPS

SHRINKAGE REINFORCEMENT

BOTTOM BEAM REINFORCEMENT

CHANNEL

ANGLE

ANCHOR BOLTS A307

TYPICAL DETAIL FOR CONNECTION BETWEEN
STEEL LANDING STAIR AND CONCRETE BEAM

LANDING STAIR CONNECTION
CHECK LIST
CHANNEL
METAL DECK
ANGLE
CONCRETE BEAM
BOTTOM BEAM REINFORCEMENT
TOP BEAM REINFORCEMENT
SHRINKAGE REINFORCEMENT
STIRRUPS REINFORCEMENT
CONCRETE SLAB
TOP SLAB REINFORCEMENT
BOTTOM SLAB REINFORCEMENT
ANCHOR BOLTS
WELDING

CHANNEL

CONCRETE BEAM

CHANNEL

PLAN VIEW

STAIR8

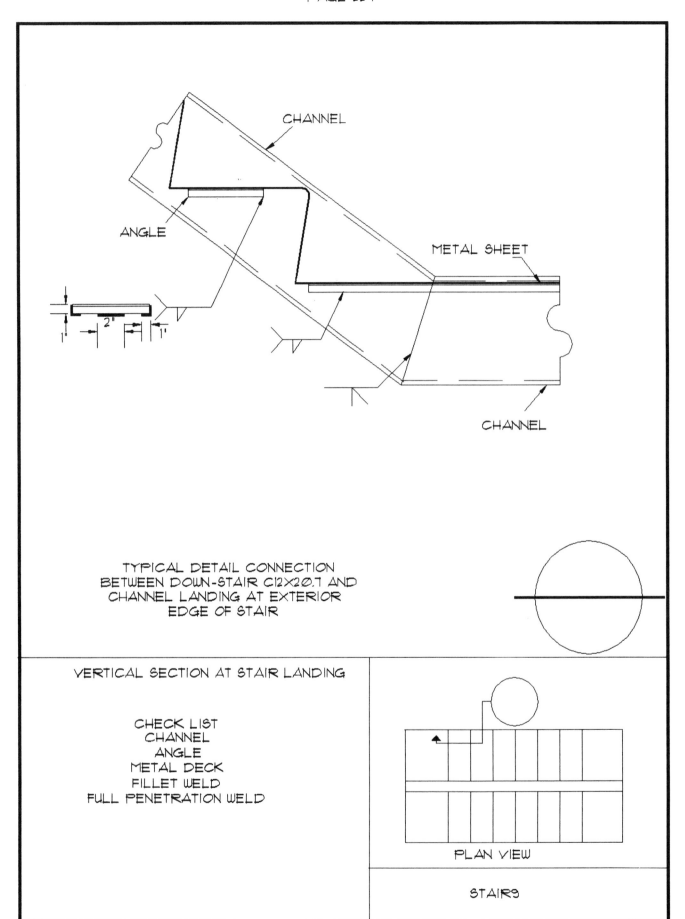

CHANNEL

ANGLE

METAL SHEET

2"

1"

1"

CHANNEL

TYPICAL DETAIL CONNECTION
BETWEEN DOWN-STAIR C12X20.7 AND
CHANNEL LANDING AT EXTERIOR
EDGE OF STAIR

VERTICAL SECTION AT STAIR LANDING

CHECK LIST
CHANNEL
ANGLE
METAL DECK
FILLET WELD
FULL PENETRATION WELD

PLAN VIEW

STAIRS

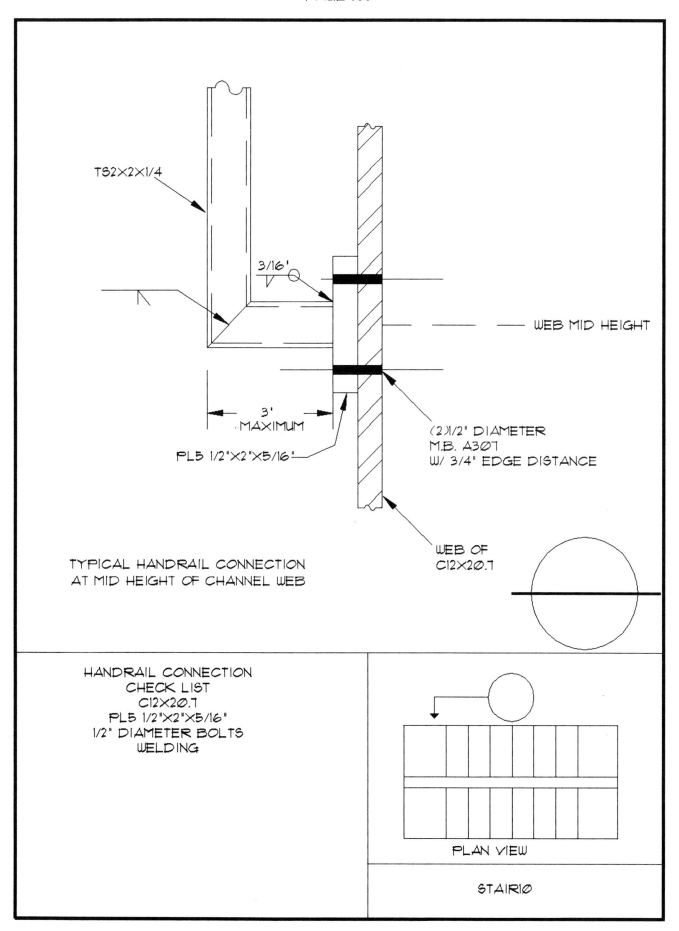

TS2X2X1/4

3/16'

WEB MID HEIGHT

3'
MAXIMUM

PL5 1/2"X2"X5/16"

(2)1/2" DIAMETER
M.B. A307
W/ 3/4" EDGE DISTANCE

WEB OF
C12X20.7

TYPICAL HANDRAIL CONNECTION
AT MID HEIGHT OF CHANNEL WEB

HANDRAIL CONNECTION
CHECK LIST
C12X20.7
PL5 1/2"X2"X5/16"
1/2" DIAMETER BOLTS
WELDING

PLAN VIEW

STAIR10

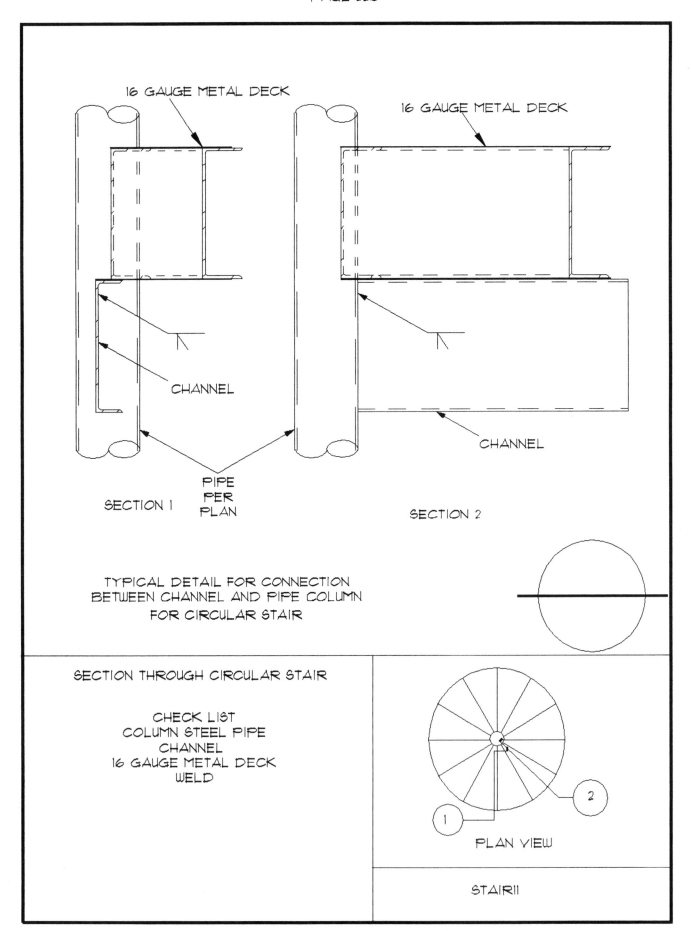

16 GAUGE METAL DECK

16 GAUGE METAL DECK

CHANNEL

CHANNEL

PIPE
PER
PLAN

SECTION 1

SECTION 2

TYPICAL DETAIL FOR CONNECTION
BETWEEN CHANNEL AND PIPE COLUMN
FOR CIRCULAR STAIR

SECTION THROUGH CIRCULAR STAIR

CHECK LIST
COLUMN STEEL PIPE
CHANNEL
16 GAUGE METAL DECK
WELD

PLAN VIEW

STAIR11

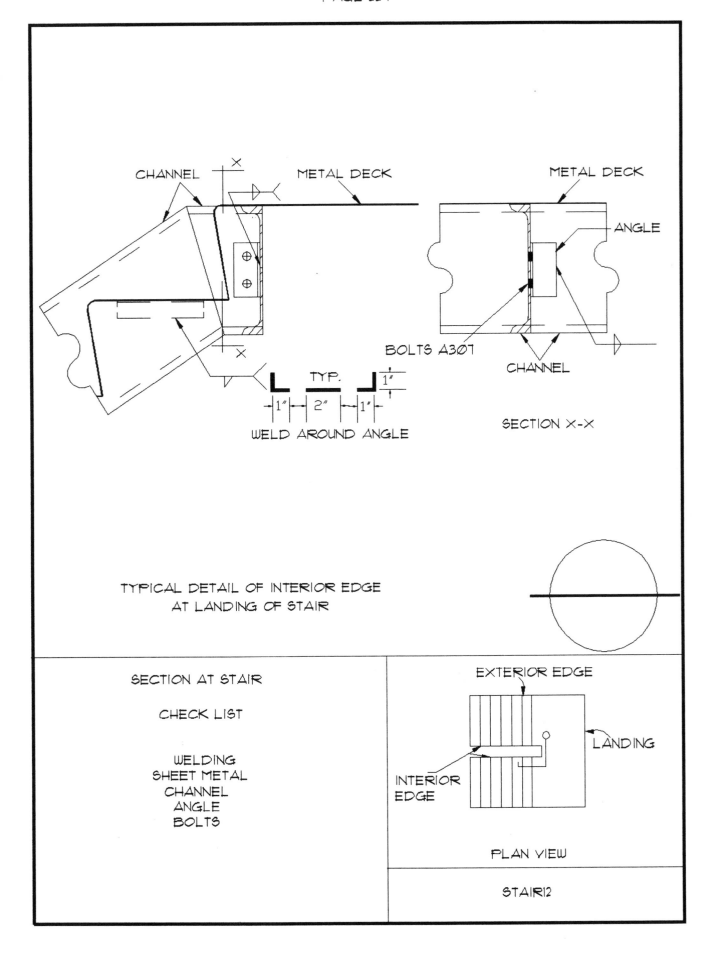

CHANNEL

METAL DECK

X

TYP.

1"

1" 2" 1"

WELD AROUND ANGLE

METAL DECK

ANGLE

BOLTS A307

CHANNEL

SECTION X-X

TYPICAL DETAIL OF INTERIOR EDGE
AT LANDING OF STAIR

SECTION AT STAIR

CHECK LIST

WELDING
SHEET METAL
CHANNEL
ANGLE
BOLTS

EXTERIOR EDGE

LANDING

INTERIOR
EDGE

PLAN VIEW

STAIR12

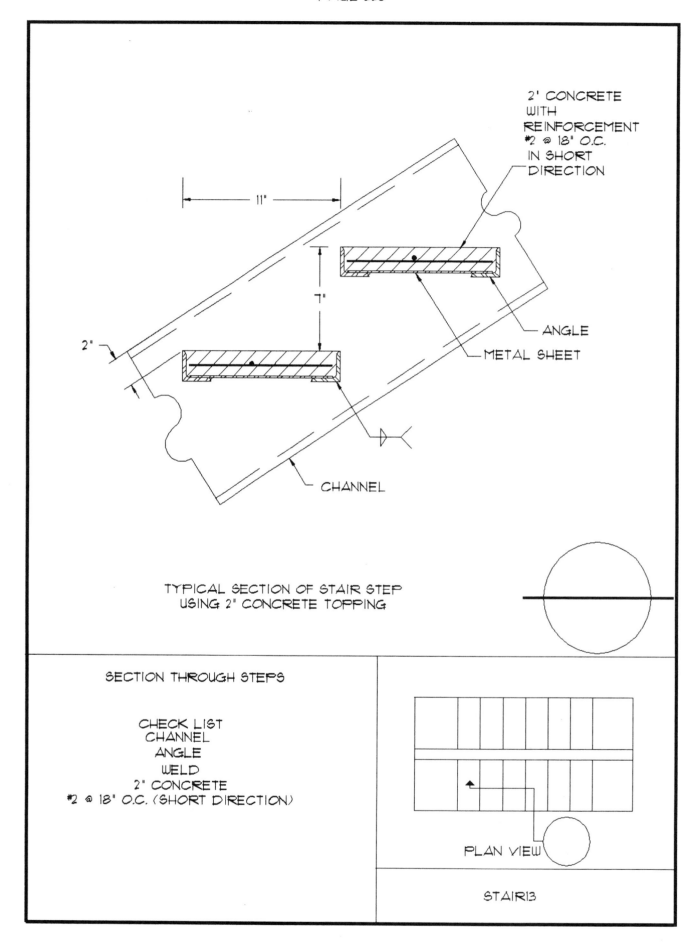

2' CONCRETE
WITH
REINFORCEMENT
#2 @ 18" O.C.
IN SHORT
DIRECTION

11"

7"

2"

ANGLE

METAL SHEET

CHANNEL

TYPICAL SECTION OF STAIR STEP
USING 2" CONCRETE TOPPING

SECTION THROUGH STEPS

CHECK LIST
CHANNEL
ANGLE
WELD
2" CONCRETE
#2 @ 18" O.C. (SHORT DIRECTION)

PLAN VIEW

STAIR13

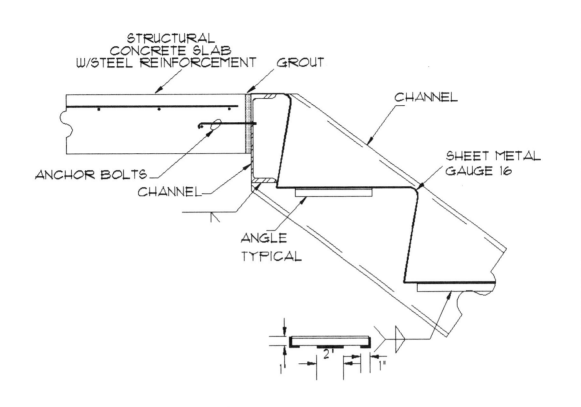

STRUCTURAL
CONCRETE SLAB
W/STEEL REINFORCEMENT   GROUT

CHANNEL

SHEET METAL
GAUGE 16

ANCHOR BOLTS

CHANNEL

ANGLE
TYPICAL

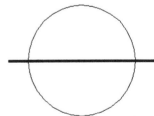

2'   1"   1'

TYPICAL DETAIL FOR STRINGER
CONNECTION WITH CONCRETE STRUCTURAL SLAB

STAIR-CONCRETE CONNECTION
CHECK LIST
WELDING
CONCRETE STRUCTURAL SLAB
ANCHOR BOLTS
SHEET METAL
ANGLE
CHANNEL

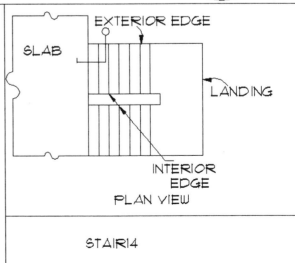

EXTERIOR EDGE

SLAB

LANDING

INTERIOR
EDGE
PLAN VIEW

STAIR14

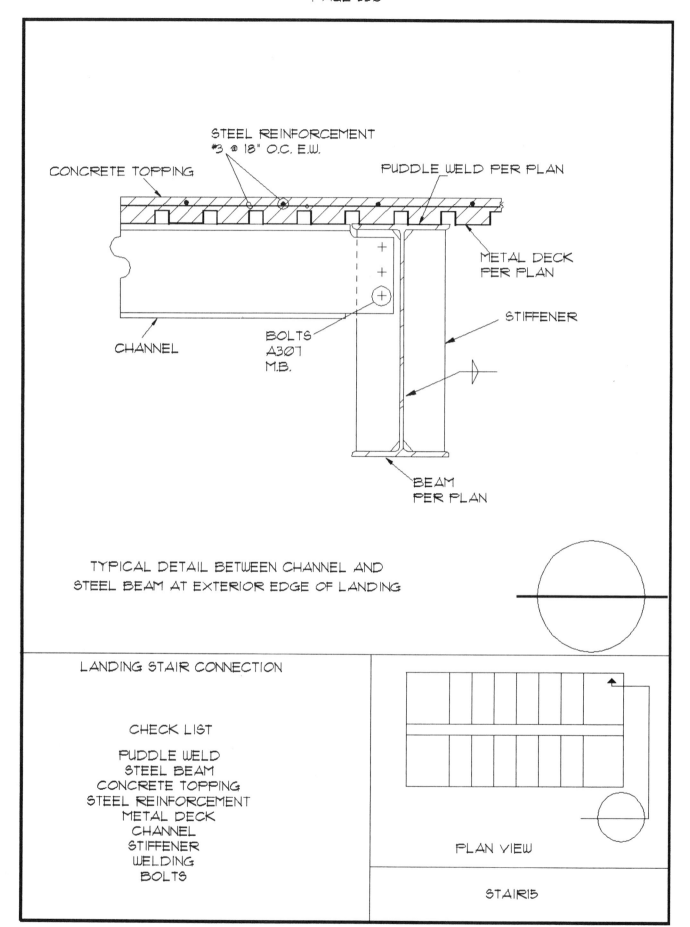

STEEL REINFORCEMENT
#3 @ 18" O.C. E.W.

CONCRETE TOPPING

PUDDLE WELD PER PLAN

METAL DECK
PER PLAN

STIFFENER

CHANNEL

BOLTS
A307
M.B.

BEAM
PER PLAN

TYPICAL DETAIL BETWEEN CHANNEL AND
STEEL BEAM AT EXTERIOR EDGE OF LANDING

LANDING STAIR CONNECTION

CHECK LIST

PUDDLE WELD
STEEL BEAM
CONCRETE TOPPING
STEEL REINFORCEMENT
METAL DECK
CHANNEL
STIFFENER
WELDING
BOLTS

PLAN VIEW

STAIRS

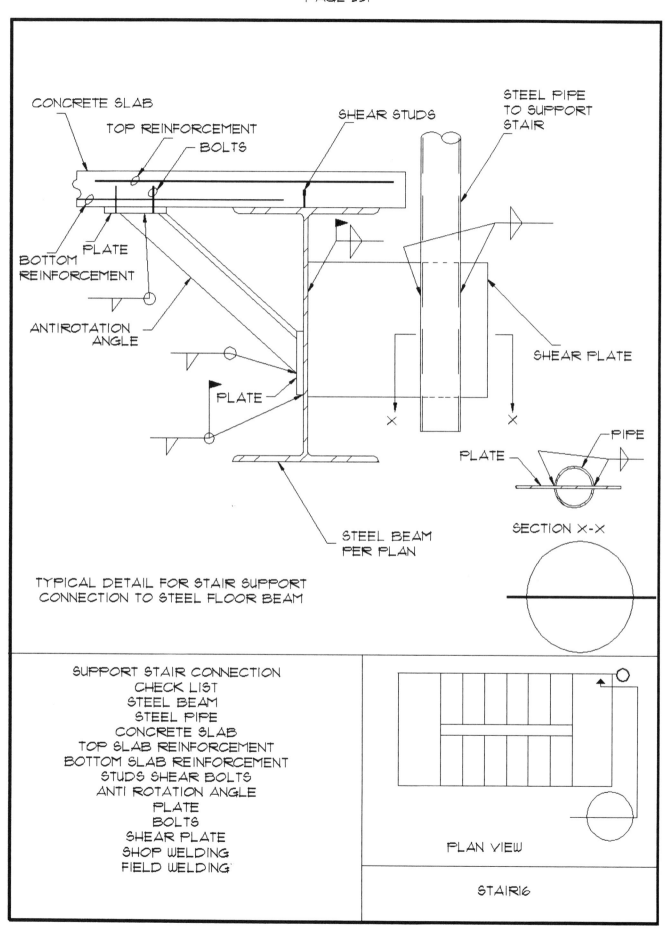

CONCRETE SLAB

TOP REINFORCEMENT

BOLTS

SHEAR STUDS

STEEL PIPE TO SUPPORT STAIR

BOTTOM REINFORCEMENT

PLATE

ANTIROTATION ANGLE

SHEAR PLATE

PLATE

STEEL BEAM PER PLAN

PLATE

PIPE

SECTION X-X

TYPICAL DETAIL FOR STAIR SUPPORT CONNECTION TO STEEL FLOOR BEAM

SUPPORT STAIR CONNECTION
CHECK LIST
STEEL BEAM
STEEL PIPE
CONCRETE SLAB
TOP SLAB REINFORCEMENT
BOTTOM SLAB REINFORCEMENT
STUDS SHEAR BOLTS
ANTI ROTATION ANGLE
PLATE
BOLTS
SHEAR PLATE
SHOP WELDING
FIELD WELDING

PLAN VIEW

STAIR16

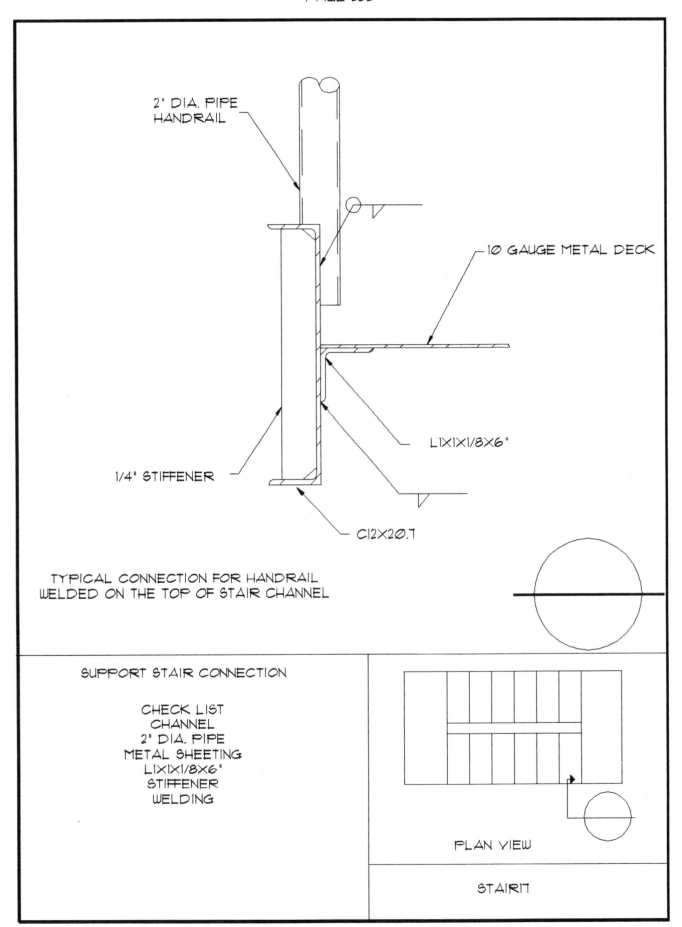

2' DIA. PIPE
HANDRAIL

10 GAUGE METAL DECK

L1X1X1/8X6"

1/4" STIFFENER

C12X20.7

TYPICAL CONNECTION FOR HANDRAIL
WELDED ON THE TOP OF STAIR CHANNEL

SUPPORT STAIR CONNECTION

CHECK LIST
CHANNEL
2" DIA. PIPE
METAL SHEETING
L1X1X1/8X6"
STIFFENER
WELDING

PLAN VIEW

STAIR17

CHAPTER 16

METAL STUDS
CONNECTION

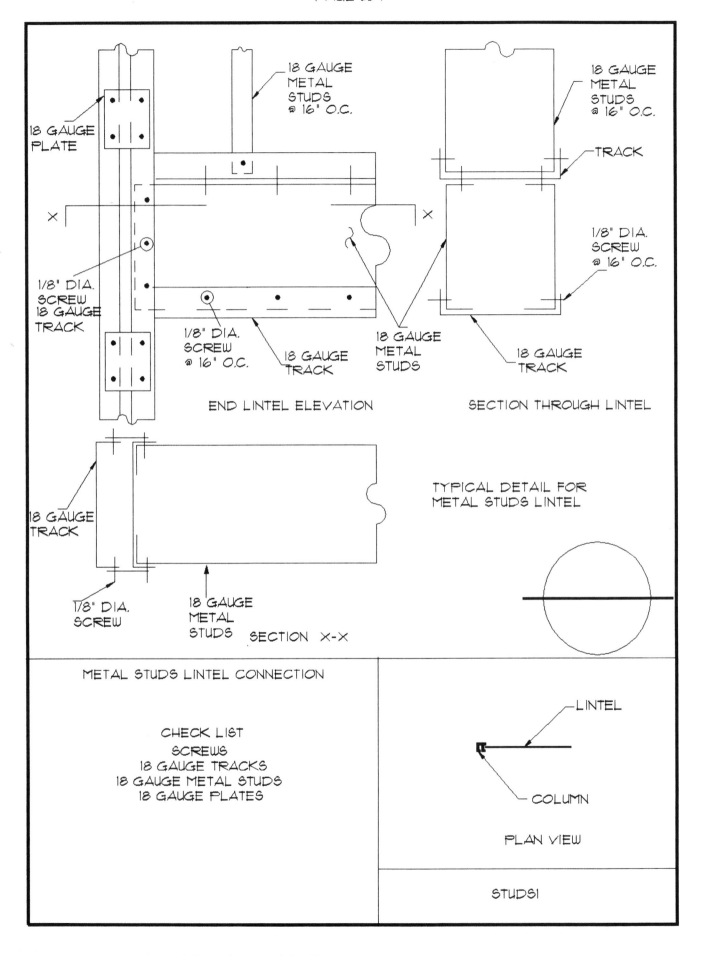

18 GAUGE PLATE

18 GAUGE METAL STUDS @ 16" O.C.

1/8" DIA. SCREW 18 GAUGE TRACK

1/8" DIA. SCREW @ 16" O.C.

18 GAUGE TRACK

18 GAUGE METAL STUDS

END LINTEL ELEVATION

18 GAUGE METAL STUDS @ 16" O.C.

TRACK

1/8" DIA. SCREW @ 16" O.C.

18 GAUGE TRACK

SECTION THROUGH LINTEL

18 GAUGE TRACK

TYPICAL DETAIL FOR METAL STUDS LINTEL

1/8" DIA. SCREW

18 GAUGE METAL STUDS

SECTION X-X

METAL STUDS LINTEL CONNECTION

CHECK LIST
SCREWS
18 GAUGE TRACKS
18 GAUGE METAL STUDS
18 GAUGE PLATES

LINTEL

COLUMN

PLAN VIEW

STUDS1

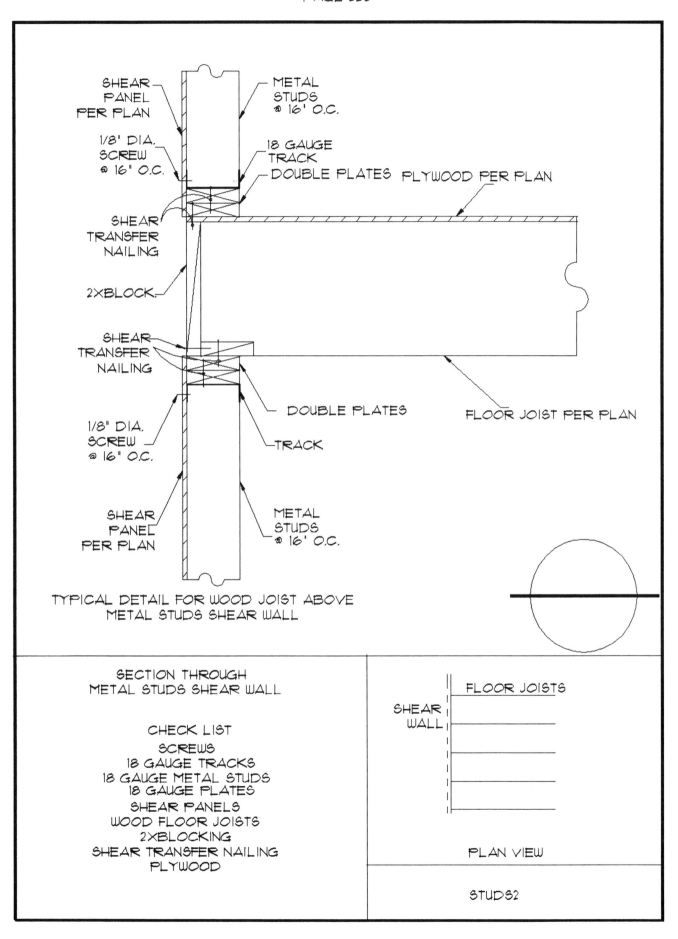

SHEAR PANEL PER PLAN

METAL STUDS @ 16' O.C.

1/8' DIA. SCREW @ 16" O.C.

18 GAUGE TRACK

DOUBLE PLATES   PLYWOOD PER PLAN

SHEAR TRANSFER NAILING

2XBLOCK

SHEAR TRANSFER NAILING

DOUBLE PLATES

1/8" DIA. SCREW @ 16" O.C.

TRACK

FLOOR JOIST PER PLAN

SHEAR PANEL PER PLAN

METAL STUDS @ 16' O.C.

TYPICAL DETAIL FOR WOOD JOIST ABOVE
METAL STUDS SHEAR WALL

SECTION THROUGH
METAL STUDS SHEAR WALL

CHECK LIST

SCREWS
18 GAUGE TRACKS
18 GAUGE METAL STUDS
18 GAUGE PLATES
SHEAR PANELS
WOOD FLOOR JOISTS
2XBLOCKING
SHEAR TRANSFER NAILING
PLYWOOD

SHEAR WALL

FLOOR JOISTS

PLAN VIEW

STUDS2

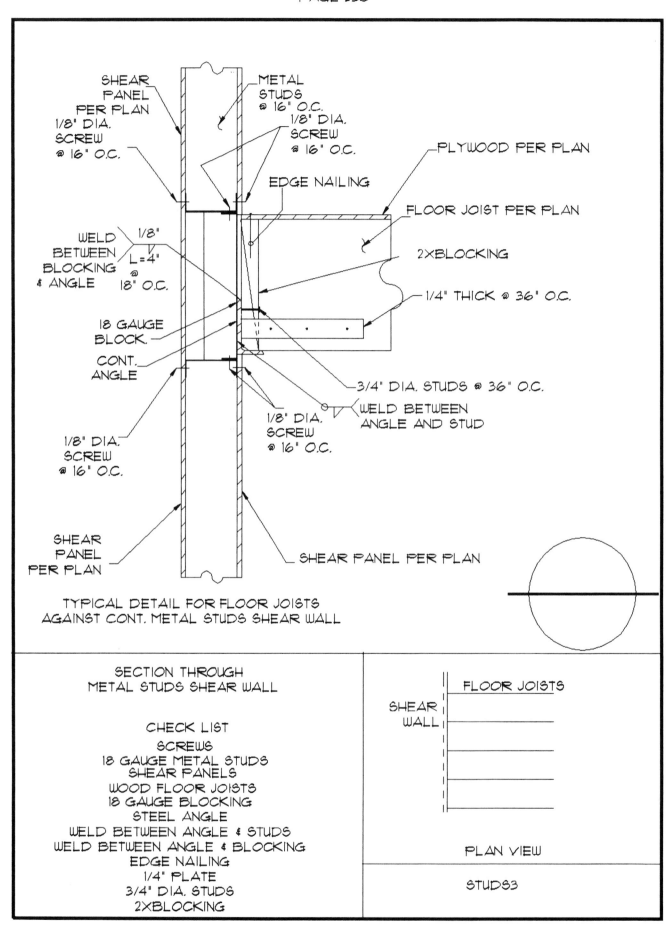

SHEAR PANEL PER PLAN 1/8" DIA. SCREW @ 16" O.C.

METAL STUDS @ 16" O.C. 1/8" DIA. SCREW @ 16" O.C.

EDGE NAILING

PLYWOOD PER PLAN

FLOOR JOIST PER PLAN

2XBLOCKING

WELD BETWEEN BLOCKING & ANGLE

1/8"
L=4"
@ 18" O.C.

1/4" THICK @ 36" O.C.

18 GAUGE BLOCK.

CONT. ANGLE

3/4" DIA. STUDS @ 36" O.C.

1/8" DIA. SCREW @ 16" O.C.

WELD BETWEEN ANGLE AND STUD

1/8" DIA. SCREW @ 16" O.C.

SHEAR PANEL PER PLAN

SHEAR PANEL PER PLAN

TYPICAL DETAIL FOR FLOOR JOISTS
AGAINST CONT. METAL STUDS SHEAR WALL

SECTION THROUGH
METAL STUDS SHEAR WALL

CHECK LIST
SCREWS
18 GAUGE METAL STUDS
SHEAR PANELS
WOOD FLOOR JOISTS
18 GAUGE BLOCKING
STEEL ANGLE
WELD BETWEEN ANGLE & STUDS
WELD BETWEEN ANGLE & BLOCKING
EDGE NAILING
1/4" PLATE
3/4" DIA. STUDS
2XBLOCKING

FLOOR JOISTS

SHEAR WALL

PLAN VIEW

STUDS3

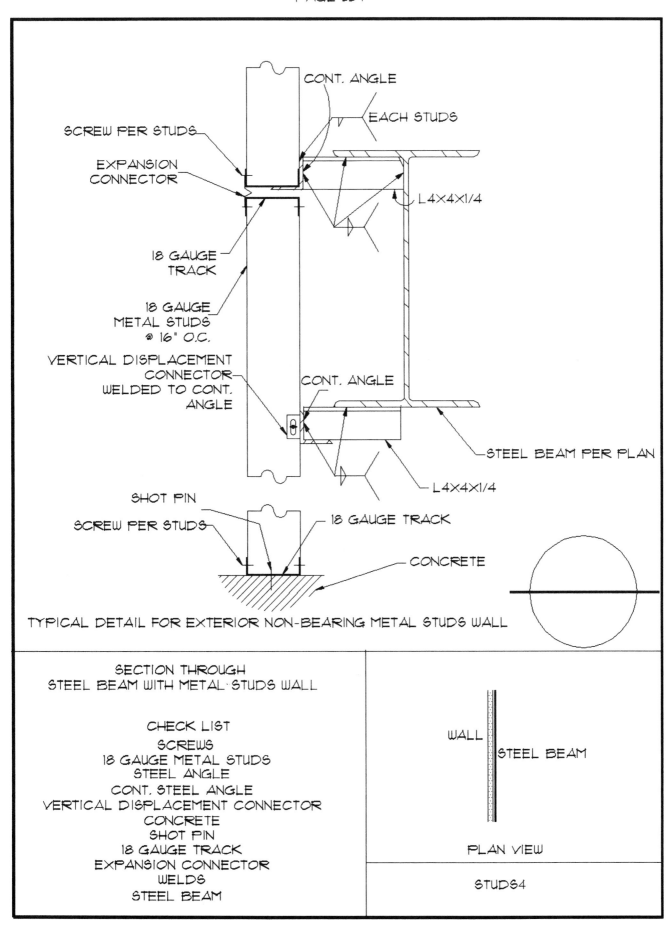

CONT. ANGLE

EACH STUDS

SCREW PER STUDS

EXPANSION CONNECTOR

L4X4X1/4

18 GAUGE TRACK

18 GAUGE METAL STUDS @ 16" O.C.

VERTICAL DISPLACEMENT CONNECTOR WELDED TO CONT. ANGLE

CONT. ANGLE

STEEL BEAM PER PLAN

L4X4X1/4

SHOT PIN

SCREW PER STUDS

18 GAUGE TRACK

CONCRETE

TYPICAL DETAIL FOR EXTERIOR NON-BEARING METAL STUDS WALL

SECTION THROUGH
STEEL BEAM WITH METAL STUDS WALL

CHECK LIST
SCREWS
18 GAUGE METAL STUDS
STEEL ANGLE
CONT. STEEL ANGLE
VERTICAL DISPLACEMENT CONNECTOR
CONCRETE
SHOT PIN
18 GAUGE TRACK
EXPANSION CONNECTOR
WELDS
STEEL BEAM

WALL

STEEL BEAM

PLAN VIEW

STUDS4

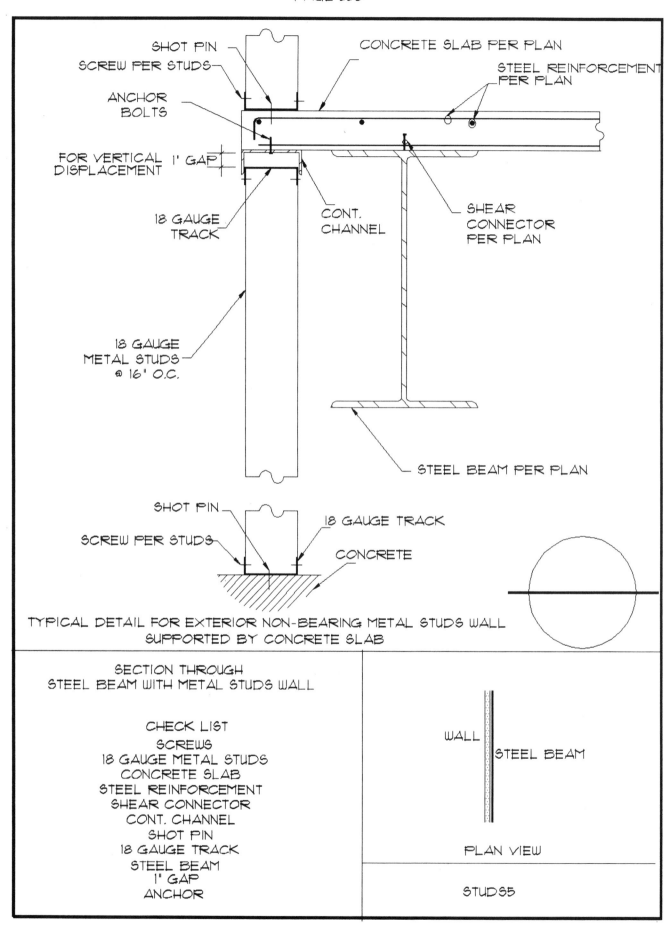

SHOT PIN

SCREW PER STUDS

ANCHOR BOLTS

FOR VERTICAL DISPLACEMENT

1' GAP

18 GAUGE TRACK

18 GAUGE METAL STUDS @ 16" O.C.

CONCRETE SLAB PER PLAN

STEEL REINFORCEMENT PER PLAN

CONT. CHANNEL

SHEAR CONNECTOR PER PLAN

STEEL BEAM PER PLAN

SHOT PIN

SCREW PER STUDS

18 GAUGE TRACK

CONCRETE

TYPICAL DETAIL FOR EXTERIOR NON-BEARING METAL STUDS WALL SUPPORTED BY CONCRETE SLAB

SECTION THROUGH STEEL BEAM WITH METAL STUDS WALL

CHECK LIST
SCREWS
18 GAUGE METAL STUDS
CONCRETE SLAB
STEEL REINFORCEMENT
SHEAR CONNECTOR
CONT. CHANNEL
SHOT PIN
18 GAUGE TRACK
STEEL BEAM
1" GAP
ANCHOR

WALL   STEEL BEAM

PLAN VIEW

STUDS5

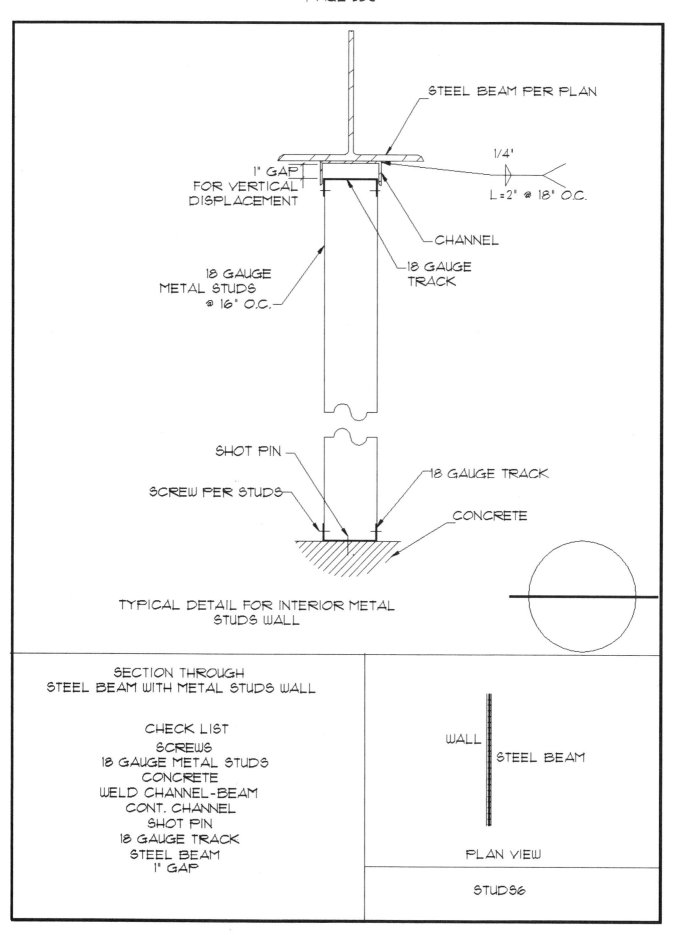

STEEL BEAM PER PLAN

1" GAP
FOR VERTICAL
DISPLACEMENT

1/4'

L=2" @ 18" O.C.

CHANNEL

18 GAUGE METAL STUDS @ 16" O.C.

18 GAUGE TRACK

SHOT PIN

18 GAUGE TRACK

SCREW PER STUDS

CONCRETE

TYPICAL DETAIL FOR INTERIOR METAL
STUDS WALL

SECTION THROUGH
STEEL BEAM WITH METAL STUDS WALL

CHECK LIST
SCREWS
18 GAUGE METAL STUDS
CONCRETE
WELD CHANNEL-BEAM
CONT. CHANNEL
SHOT PIN
18 GAUGE TRACK
STEEL BEAM
1" GAP

WALL   STEEL BEAM

PLAN VIEW

STUDS6

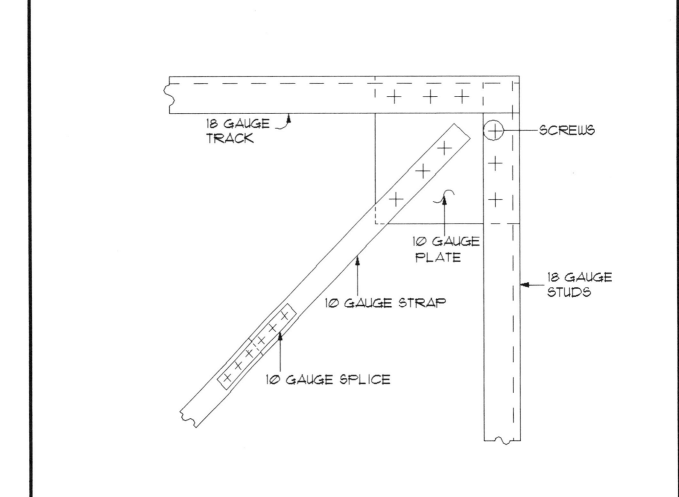

18 GAUGE
TRACK

SCREWS

10 GAUGE
PLATE

18 GAUGE
STUDS

10 GAUGE STRAP

10 GAUGE SPLICE

TYPICAL DETAIL FOR DIAGONAL STRAP
CONNECTED TO THE CORNER OF METAL STUDS WALL

ELEVATION OF METAL STUDS WALL

CHECK LIST
SCREWS
18 GAUGE METAL STUDS
10 GAUGE STRAP
18 GAUGE SPLICE
18 GAUGE TRACK

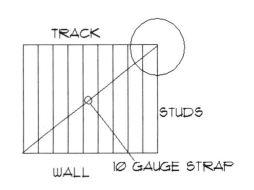

TRACK

STUDS

WALL

10 GAUGE STRAP

STUDS7

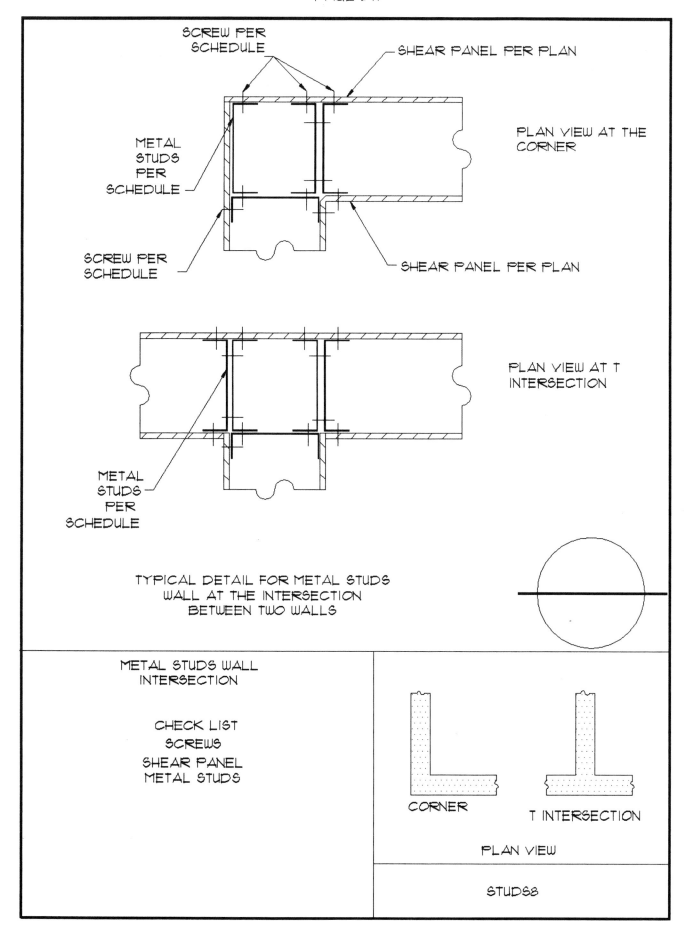

SCREW PER SCHEDULE

SHEAR PANEL PER PLAN

PLAN VIEW AT THE CORNER

METAL STUDS PER SCHEDULE

SCREW PER SCHEDULE

SHEAR PANEL PER PLAN

PLAN VIEW AT T INTERSECTION

METAL STUDS PER SCHEDULE

TYPICAL DETAIL FOR METAL STUDS WALL AT THE INTERSECTION BETWEEN TWO WALLS

METAL STUDS WALL INTERSECTION

CHECK LIST
SCREWS
SHEAR PANEL
METAL STUDS

CORNER

T INTERSECTION

PLAN VIEW

STUDS8

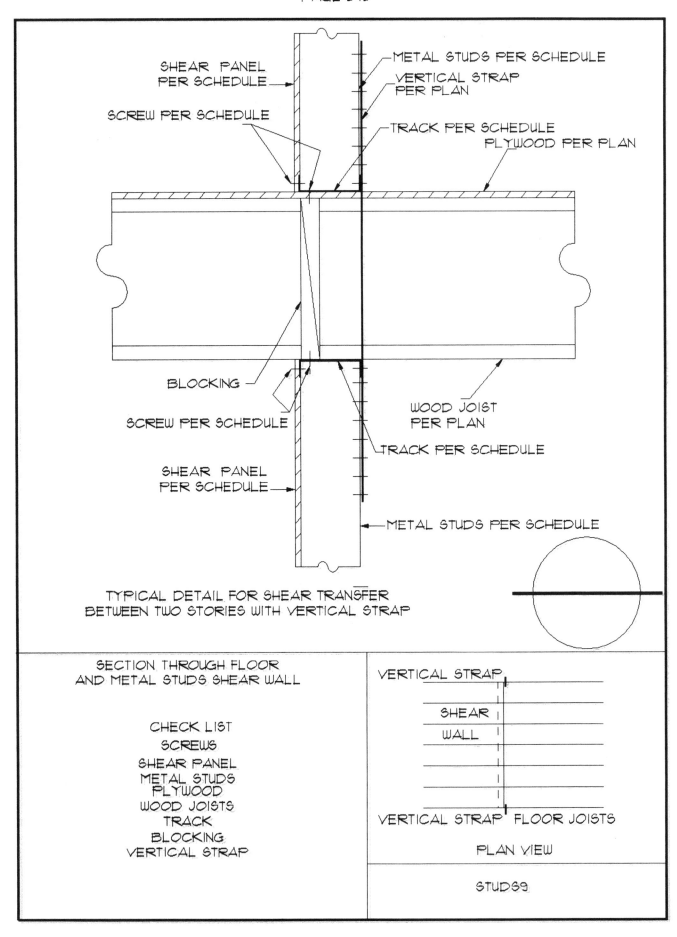

SHEAR PANEL
PER SCHEDULE

METAL STUDS PER SCHEDULE

VERTICAL STRAP
PER PLAN

SCREW PER SCHEDULE

TRACK PER SCHEDULE
PLYWOOD PER PLAN

BLOCKING

SCREW PER SCHEDULE

WOOD JOIST
PER PLAN

TRACK PER SCHEDULE

SHEAR PANEL
PER SCHEDULE

METAL STUDS PER SCHEDULE

TYPICAL DETAIL FOR SHEAR TRANSFER
BETWEEN TWO STORIES WITH VERTICAL STRAP

SECTION THROUGH FLOOR
AND METAL STUDS SHEAR WALL

CHECK LIST
SCREWS
SHEAR PANEL
METAL STUDS
PLYWOOD
WOOD JOISTS
TRACK
BLOCKING
VERTICAL STRAP

VERTICAL STRAP

SHEAR

WALL

VERTICAL STRAP FLOOR JOISTS

PLAN VIEW

STUDS9

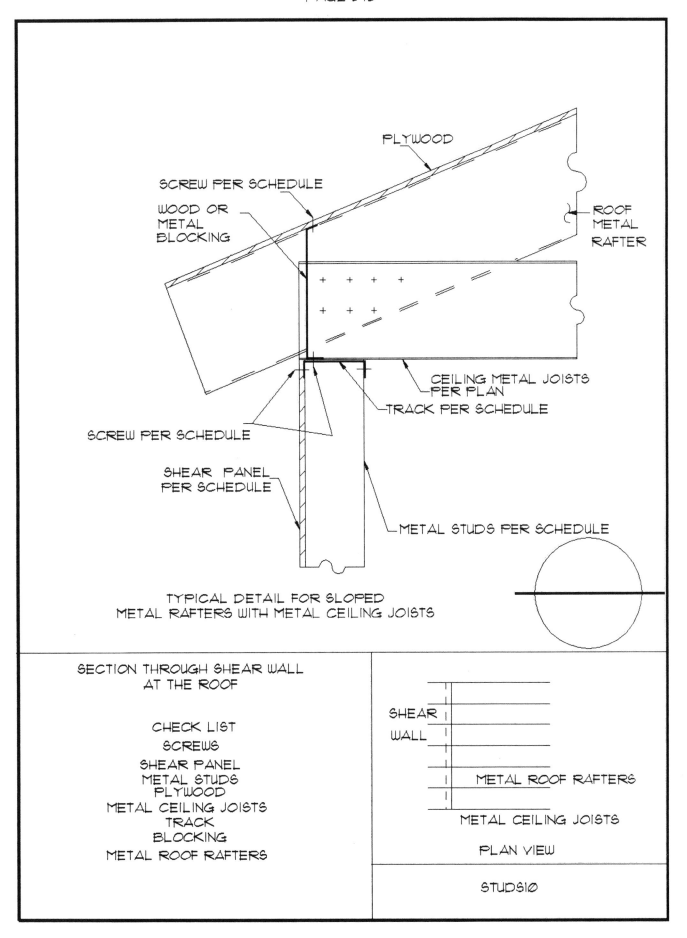

PLYWOOD

SCREW PER SCHEDULE

WOOD OR
METAL
BLOCKING

ROOF
METAL
RAFTER

CEILING METAL JOISTS
PER PLAN

TRACK PER SCHEDULE

SCREW PER SCHEDULE

SHEAR PANEL
PER SCHEDULE

METAL STUDS PER SCHEDULE

TYPICAL DETAIL FOR SLOPED
METAL RAFTERS WITH METAL CEILING JOISTS

SECTION THROUGH SHEAR WALL
AT THE ROOF

CHECK LIST
SCREWS
SHEAR PANEL
METAL STUDS
PLYWOOD
METAL CEILING JOISTS
TRACK
BLOCKING
METAL ROOF RAFTERS

SHEAR
WALL

METAL ROOF RAFTERS

METAL CEILING JOISTS

PLAN VIEW

STUDS10

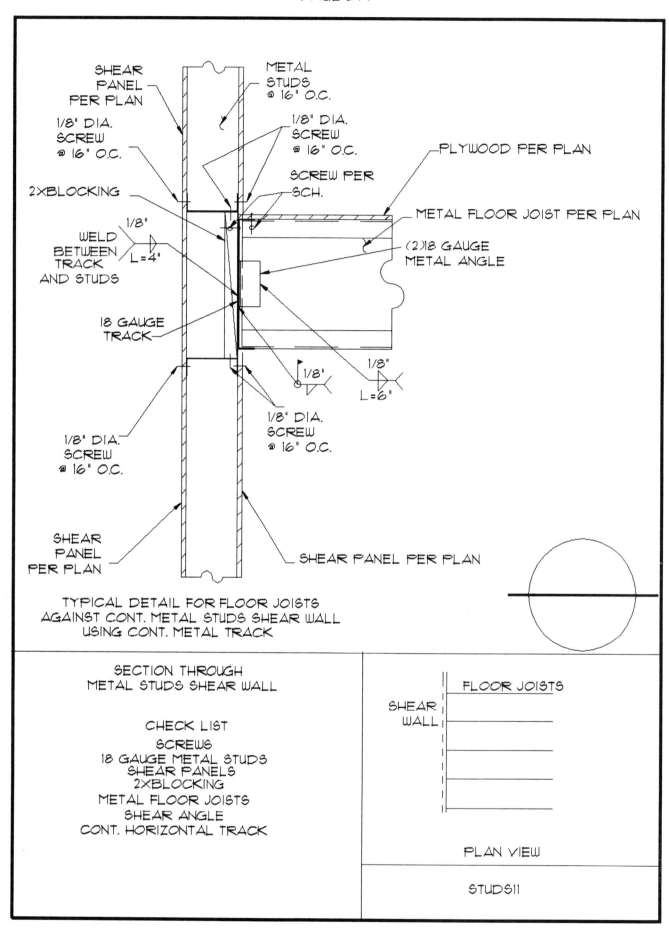

SHEAR PANEL PER PLAN

METAL STUDS @ 16" O.C.

1/8" DIA. SCREW @ 16" O.C.

1/8" DIA. SCREW @ 16" O.C.

PLYWOOD PER PLAN

SCREW PER SCH.

2XBLOCKING

METAL FLOOR JOIST PER PLAN

WELD BETWEEN TRACK AND STUDS

1/8"
L=4"

(2)18 GAUGE METAL ANGLE

18 GAUGE TRACK

1/8"

1/8"
L=6"

1/8" DIA. SCREW @ 16" O.C.

1/8" DIA. SCREW @ 16" O.C.

SHEAR PANEL PER PLAN

SHEAR PANEL PER PLAN

TYPICAL DETAIL FOR FLOOR JOISTS
AGAINST CONT. METAL STUDS SHEAR WALL
USING CONT. METAL TRACK

SECTION THROUGH
METAL STUDS SHEAR WALL

CHECK LIST
SCREWS
18 GAUGE METAL STUDS
SHEAR PANELS
2XBLOCKING
METAL FLOOR JOISTS
SHEAR ANGLE
CONT. HORIZONTAL TRACK

FLOOR JOISTS

SHEAR WALL

PLAN VIEW

STUDS11

CHAPTER 17

TENSION CONNECTION

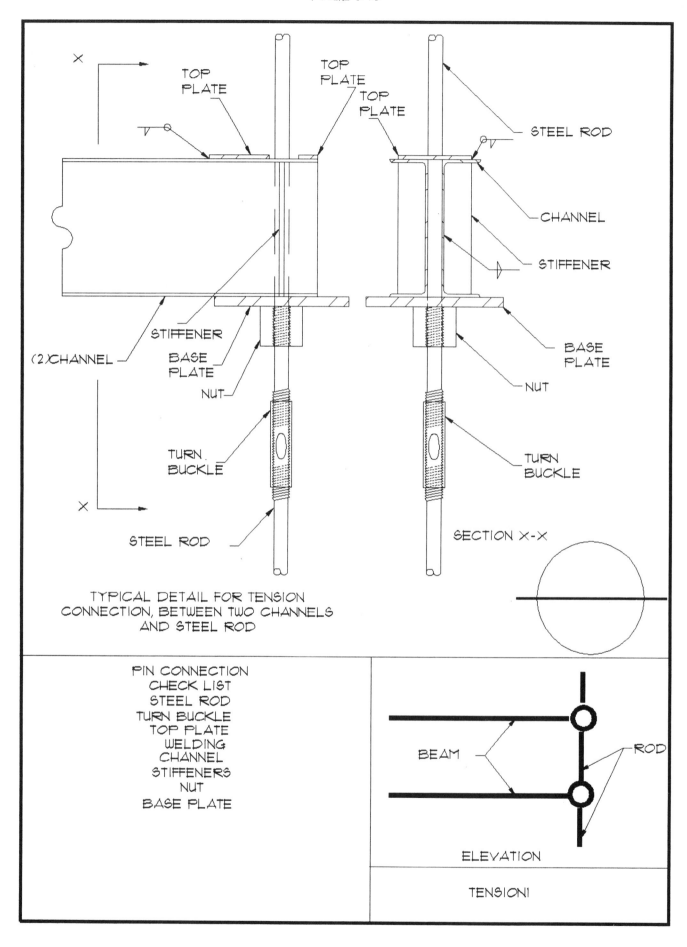

TOP PLATE

TOP PLATE

TOP PLATE

STEEL ROD

CHANNEL

STIFFENER

STIFFENER

(2)CHANNEL

BASE PLATE

NUT

BASE PLATE

NUT

TURN BUCKLE

TURN BUCKLE

STEEL ROD

SECTION X-X

TYPICAL DETAIL FOR TENSION
CONNECTION, BETWEEN TWO CHANNELS
AND STEEL ROD

PIN CONNECTION
CHECK LIST
STEEL ROD
TURN BUCKLE
TOP PLATE
WELDING
CHANNEL
STIFFENERS
NUT
BASE PLATE

BEAM

ROD

ELEVATION

TENSION1

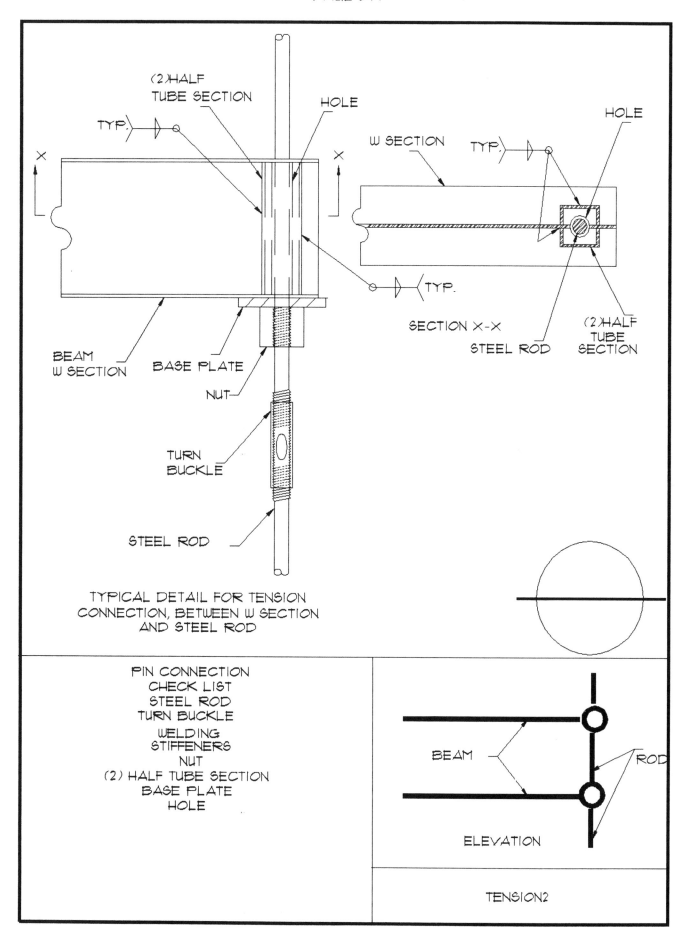

(2)HALF TUBE SECTION

HOLE

TYP.)

X

W SECTION

HOLE

TYP.)

X

TYP.

SECTION X-X

STEEL ROD

(2)HALF TUBE SECTION

BEAM W SECTION

BASE PLATE

NUT

TURN BUCKLE

STEEL ROD

TYPICAL DETAIL FOR TENSION CONNECTION, BETWEEN W SECTION AND STEEL ROD

PIN CONNECTION
CHECK LIST
STEEL ROD
TURN BUCKLE
WELDING
STIFFENERS
NUT
(2) HALF TUBE SECTION
BASE PLATE
HOLE

BEAM

ROD

ELEVATION

TENSION2

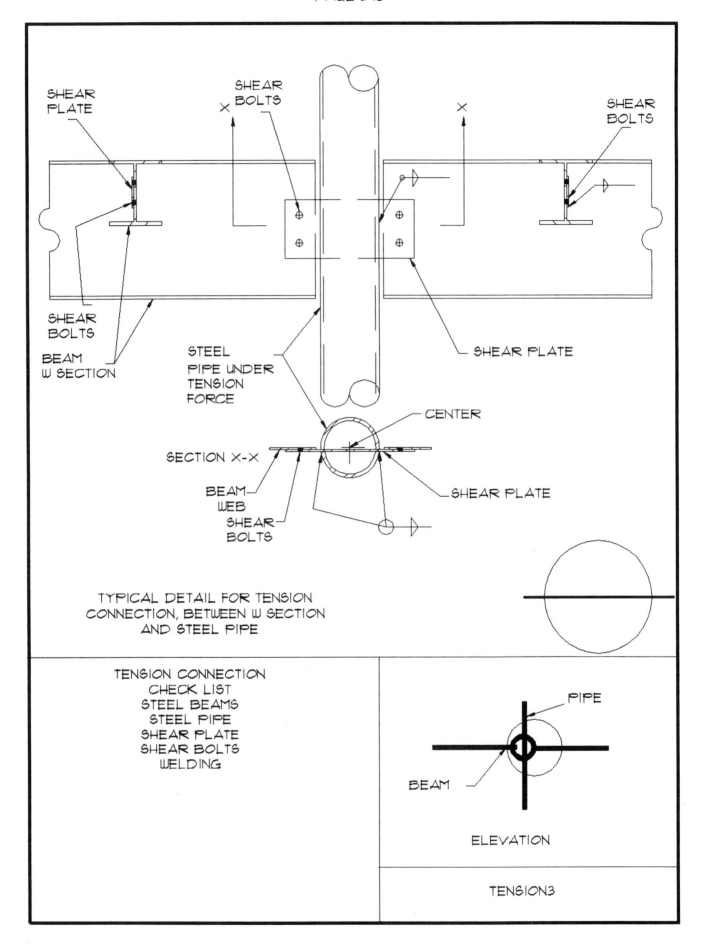

SHEAR PLATE

SHEAR BOLTS

SHEAR BOLTS

SHEAR BOLTS

BEAM W SECTION

STEEL PIPE UNDER TENSION FORCE

SHEAR PLATE

CENTER

SECTION X-X

BEAM WEB

SHEAR BOLTS

SHEAR PLATE

TYPICAL DETAIL FOR TENSION CONNECTION, BETWEEN W SECTION AND STEEL PIPE

TENSION CONNECTION
CHECK LIST
STEEL BEAMS
STEEL PIPE
SHEAR PLATE
SHEAR BOLTS
WELDING

PIPE

BEAM

ELEVATION

TENSION3

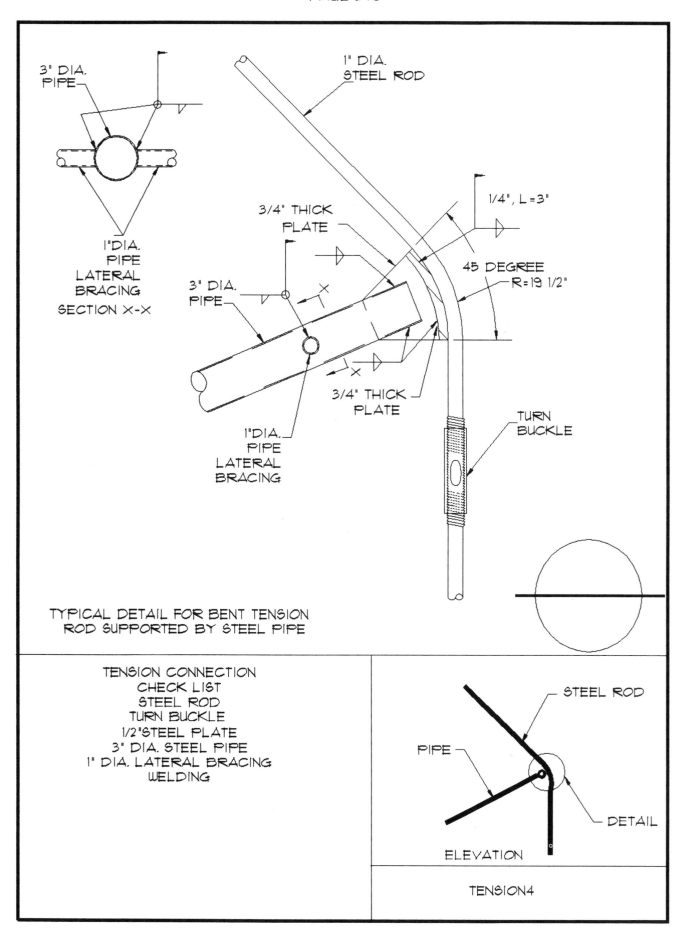

3" DIA. PIPE

1" DIA. STEEL ROD

1" DIA. PIPE LATERAL BRACING SECTION X-X

3/4" THICK PLATE

3" DIA. PIPE

1/4", L=3"

45 DEGREE
R=19 1/2"

3/4" THICK PLATE

1" DIA. PIPE LATERAL BRACING

TURN BUCKLE

TYPICAL DETAIL FOR BENT TENSION ROD SUPPORTED BY STEEL PIPE

TENSION CONNECTION
CHECK LIST
STEEL ROD
TURN BUCKLE
1/2" STEEL PLATE
3" DIA. STEEL PIPE
1" DIA. LATERAL BRACING
WELDING

STEEL ROD

PIPE

DETAIL

ELEVATION

TENSION4

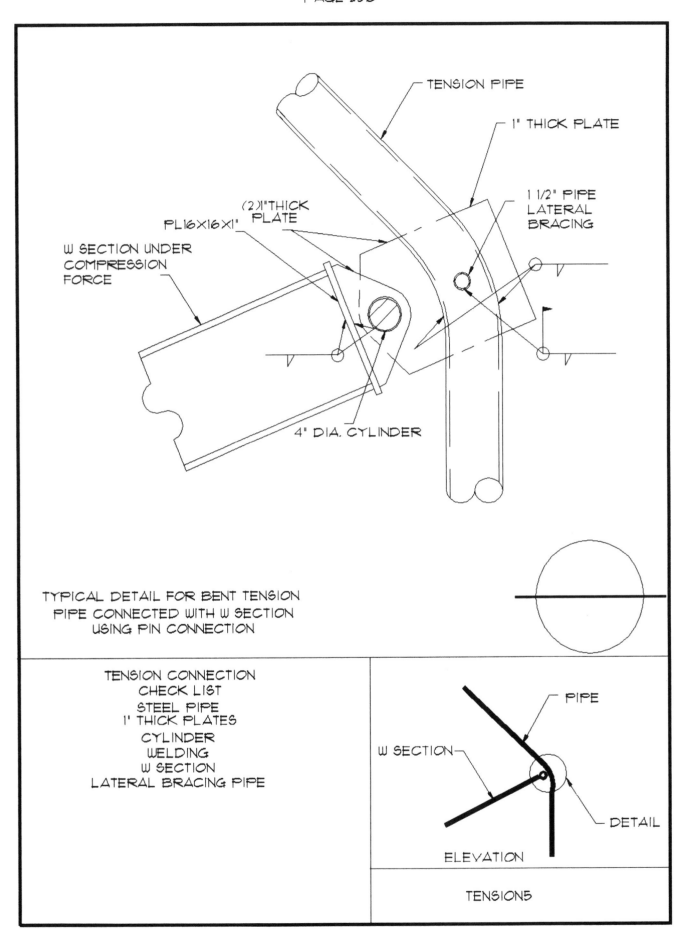

TENSION PIPE

1" THICK PLATE

1 1/2" PIPE LATERAL BRACING

(2)1"THICK PLATE

PL16X16X1"

W SECTION UNDER COMPRESSION FORCE

4" DIA. CYLINDER

TYPICAL DETAIL FOR BENT TENSION
PIPE CONNECTED WITH W SECTION
USING PIN CONNECTION

TENSION CONNECTION
CHECK LIST
STEEL PIPE
1' THICK PLATES
CYLINDER
WELDING
W SECTION
LATERAL BRACING PIPE

PIPE

W SECTION

DETAIL

ELEVATION

TENSION5

CHAPTER 18

TORSION CONNECTION

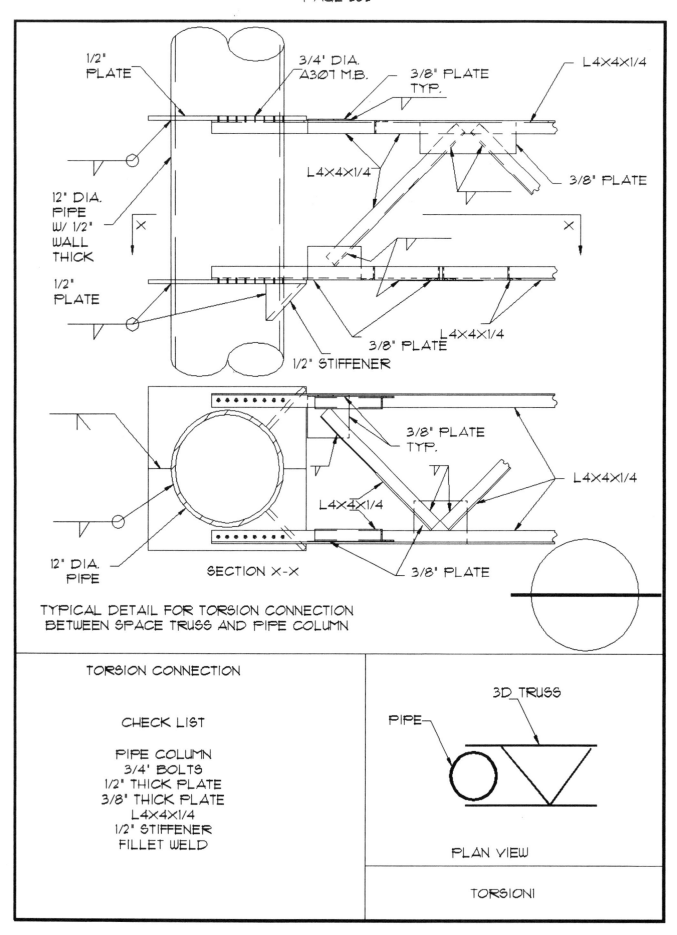

1/2" PLATE

3/4' DIA. A307 M.B.

3/8" PLATE TYP.

L4X4X1/4

12" DIA. PIPE W/ 1/2" WALL THICK

L4X4X1/4

3/8" PLATE

1/2" PLATE

3/8" PLATE

1/2" STIFFENER

L4X4X1/4

SECTION X-X

12" DIA. PIPE

3/8" PLATE TYP.

L4X4X1/4

L4X4X1/4

3/8" PLATE

TYPICAL DETAIL FOR TORSION CONNECTION
BETWEEN SPACE TRUSS AND PIPE COLUMN

TORSION CONNECTION

CHECK LIST

PIPE COLUMN
3/4' BOLTS
1/2" THICK PLATE
3/8" THICK PLATE
L4X4X1/4
1/2" STIFFENER
FILLET WELD

PIPE

3D_TRUSS

PLAN VIEW

TORSION1

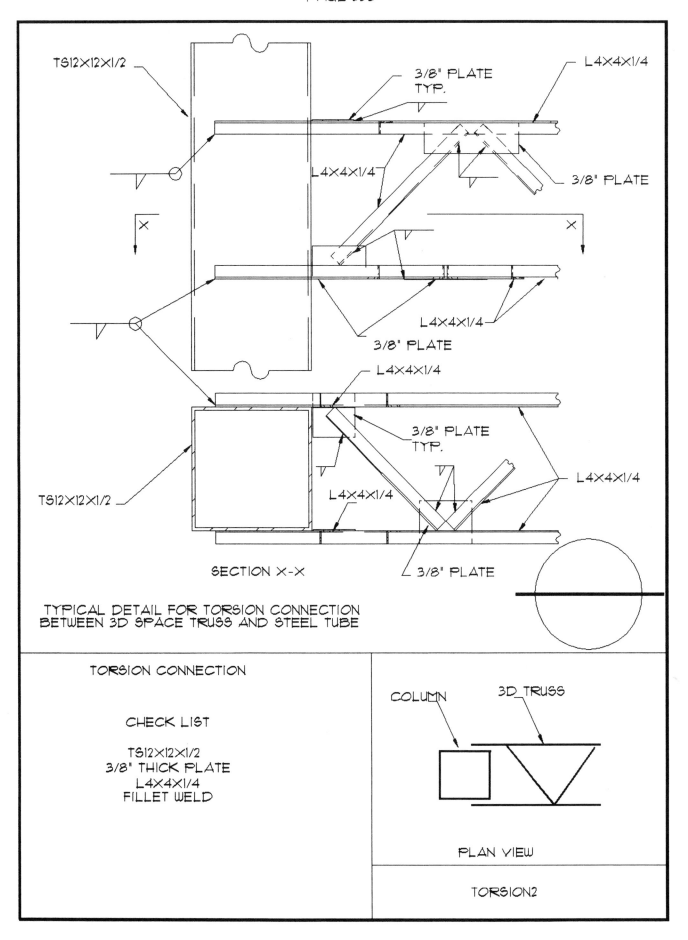

TS12X12X1/2

3/8" PLATE
TYP.

L4X4X1/4

L4X4X1/4

3/8" PLATE

L4X4X1/4

3/8" PLATE

L4X4X1/4

L4X4X1/4

3/8" PLATE
TYP.

L4X4X1/4

TS12X12X1/2

L4X4X1/4

3/8" PLATE

SECTION X-X

TYPICAL DETAIL FOR TORSION CONNECTION
BETWEEN 3D SPACE TRUSS AND STEEL TUBE

TORSION CONNECTION

CHECK LIST

TS12X12X1/2
3/8" THICK PLATE
L4X4X1/4
FILLET WELD

COLUMN   3D TRUSS

PLAN VIEW

TORSION2

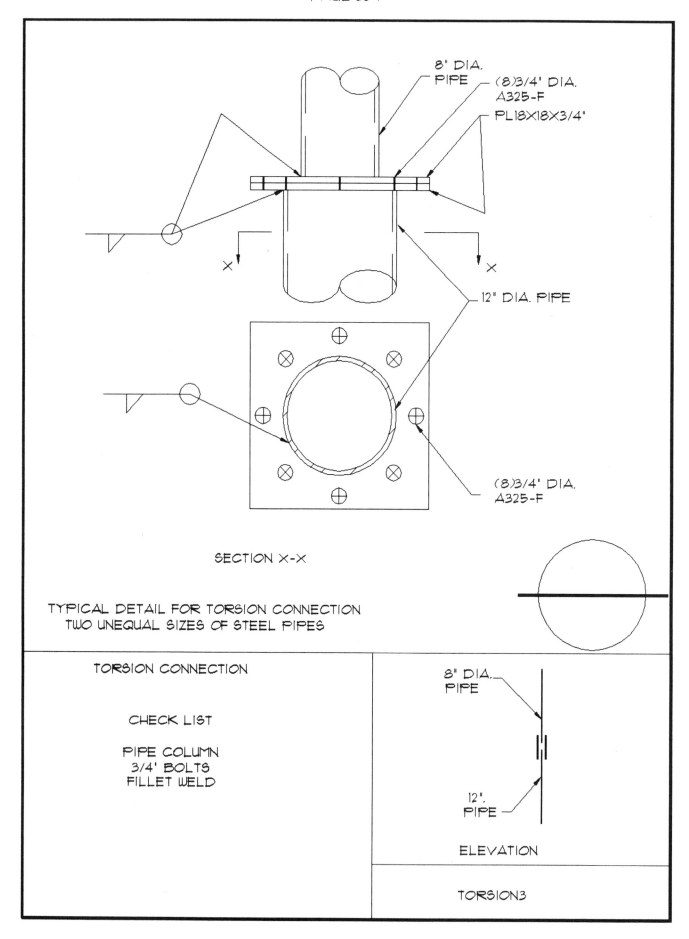

8" DIA. PIPE

(8)3/4' DIA. A325-F

PL18X18X3/4"

X X

12" DIA. PIPE

(8)3/4' DIA. A325-F

SECTION X-X

TYPICAL DETAIL FOR TORSION CONNECTION
TWO UNEQUAL SIZES OF STEEL PIPES

TORSION CONNECTION

CHECK LIST

PIPE COLUMN
3/4' BOLTS
FILLET WELD

8" DIA. PIPE

12". PIPE

ELEVATION

TORSION3

CHAPTER 19

TRUSSES

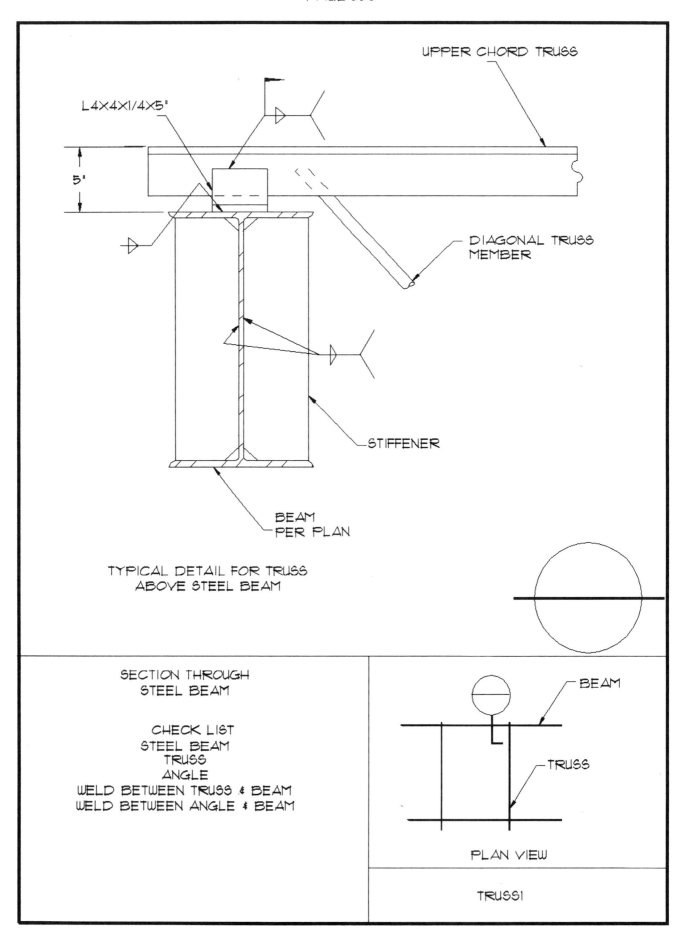

UPPER CHORD TRUSS

L4X4X1/4X5"

5"

DIAGONAL TRUSS
MEMBER

STIFFENER

BEAM
PER PLAN

TYPICAL DETAIL FOR TRUSS
ABOVE STEEL BEAM

SECTION THROUGH
STEEL BEAM

CHECK LIST
STEEL BEAM
TRUSS
ANGLE
WELD BETWEEN TRUSS & BEAM
WELD BETWEEN ANGLE & BEAM

BEAM

TRUSS

PLAN VIEW

TRUSS1

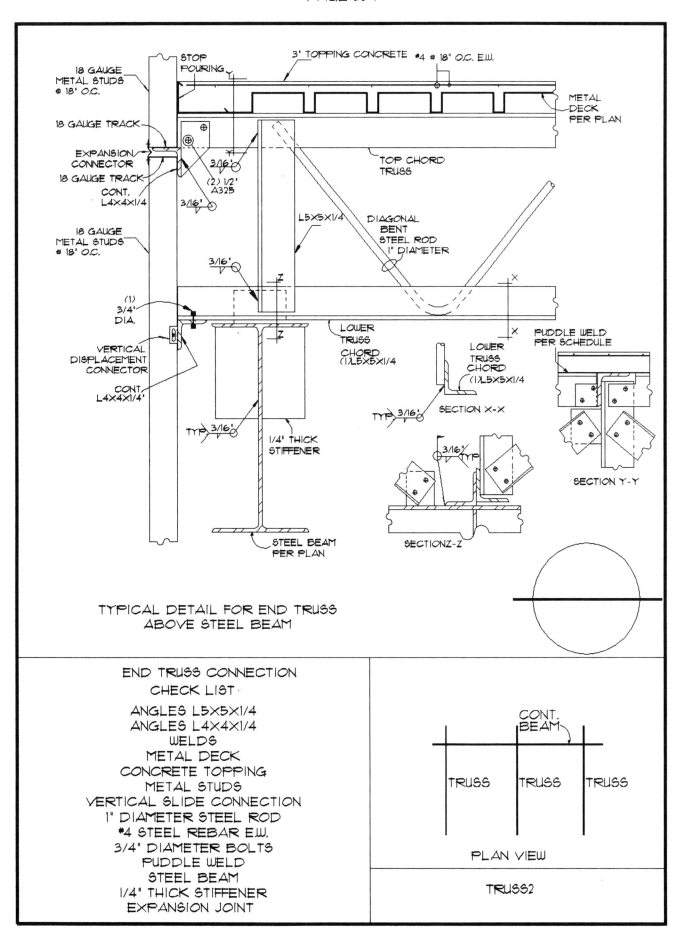

TYPICAL DETAIL FOR END TRUSS
ABOVE STEEL BEAM

END TRUSS CONNECTION
CHECK LIST

ANGLES L5X5XI/4
ANGLES L4X4XI/4
WELDS
METAL DECK
CONCRETE TOPPING
METAL STUDS
VERTICAL SLIDE CONNECTION
1" DIAMETER STEEL ROD
#4 STEEL REBAR E.W.
3/4" DIAMETER BOLTS
PUDDLE WELD
STEEL BEAM
1/4" THICK STIFFENER
EXPANSION JOINT

PLAN VIEW

TRUSS2

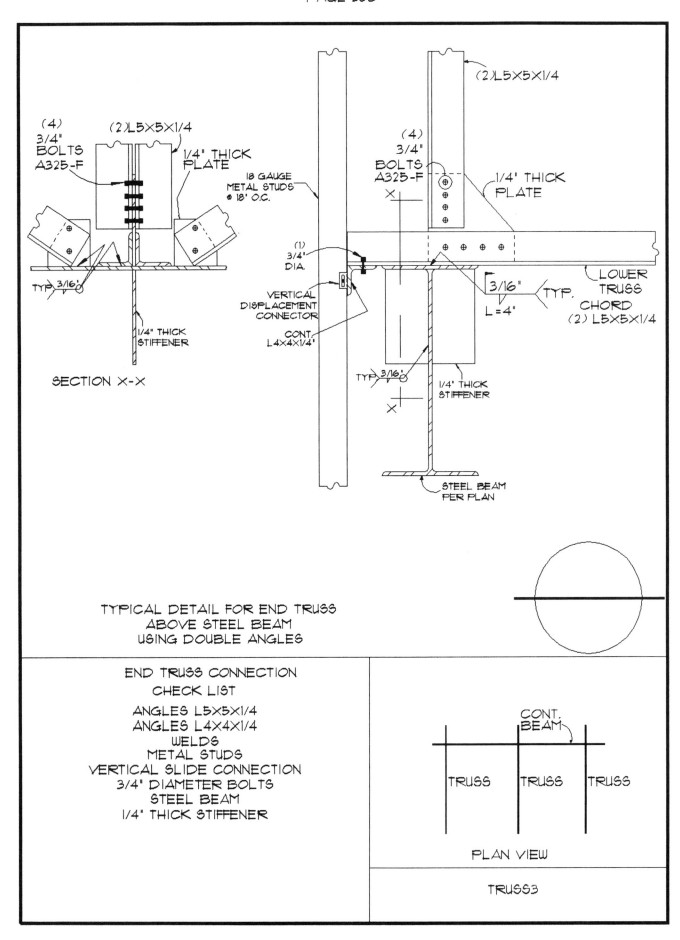

SECTION X-X

(4) 3/4" BOLTS A325-F

(2)L5X5X1/4

1/4" THICK PLATE

18 GAUGE METAL STUDS @ 18' O.C.

TYP 3/16'

1/4" THICK STIFFENER

(1) 3/4' DIA.

VERTICAL DISPLACEMENT CONNECTOR

CONT. L4X4X1/4'

(2)L5X5X1/4

(4) 3/4" BOLTS A325-F

1/4' THICK PLATE

3/16"
L = 4"

TYP.

LOWER TRUSS CHORD (2) L5X5X1/4

TYP 3/16'

1/4' THICK STIFFENER

STEEL BEAM PER PLAN

TYPICAL DETAIL FOR END TRUSS ABOVE STEEL BEAM USING DOUBLE ANGLES

END TRUSS CONNECTION CHECK LIST

ANGLES L5X5X1/4
ANGLES L4X4X1/4
WELDS
METAL STUDS
VERTICAL SLIDE CONNECTION
3/4" DIAMETER BOLTS
STEEL BEAM
1/4" THICK STIFFENER

CONT. BEAM

TRUSS   TRUSS   TRUSS

PLAN VIEW

TRUSS3

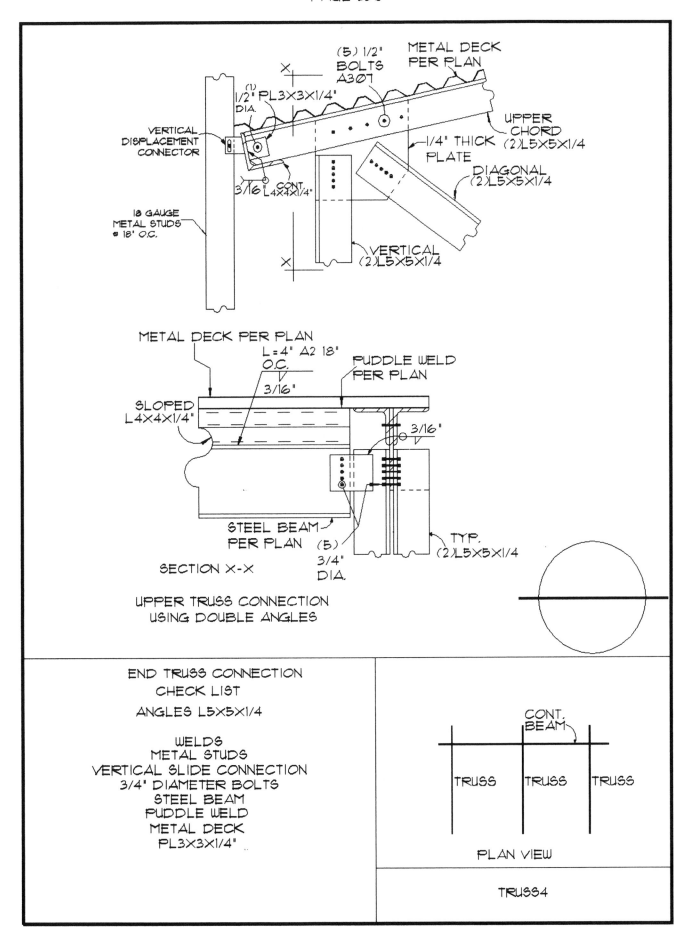

METAL DECK
PER PLAN

(5) 1/2"
BOLTS
A307

(1)
1/2" PL3X3X1/4"
DIA.

VERTICAL
DISPLACEMENT
CONNECTOR

UPPER
CHORD
(2)L5X5X1/4

1/4" THICK
PLATE

DIAGONAL
(2)L5X5X1/4

3/16" CONT.
L4X4X1/4"

18 GAUGE
METAL STUDS
@ 18' O.C.

VERTICAL
(2)L5X5X1/4

METAL DECK PER PLAN

L=4" A2 18"
O.C.

PUDDLE WELD
PER PLAN

3/16"

SLOPED
L4X4X1/4"

3/16"

STEEL BEAM
PER PLAN

(5)
3/4"
DIA.

TYP.
(2)L5X5X1/4

SECTION X-X

UPPER TRUSS CONNECTION
USING DOUBLE ANGLES

END TRUSS CONNECTION
CHECK LIST
ANGLES L5X5X1/4

WELDS
METAL STUDS
VERTICAL SLIDE CONNECTION
3/4" DIAMETER BOLTS
STEEL BEAM
PUDDLE WELD
METAL DECK
PL3X3X1/4"

CONT.
BEAM

TRUSS   TRUSS   TRUSS

PLAN VIEW

TRUSS4

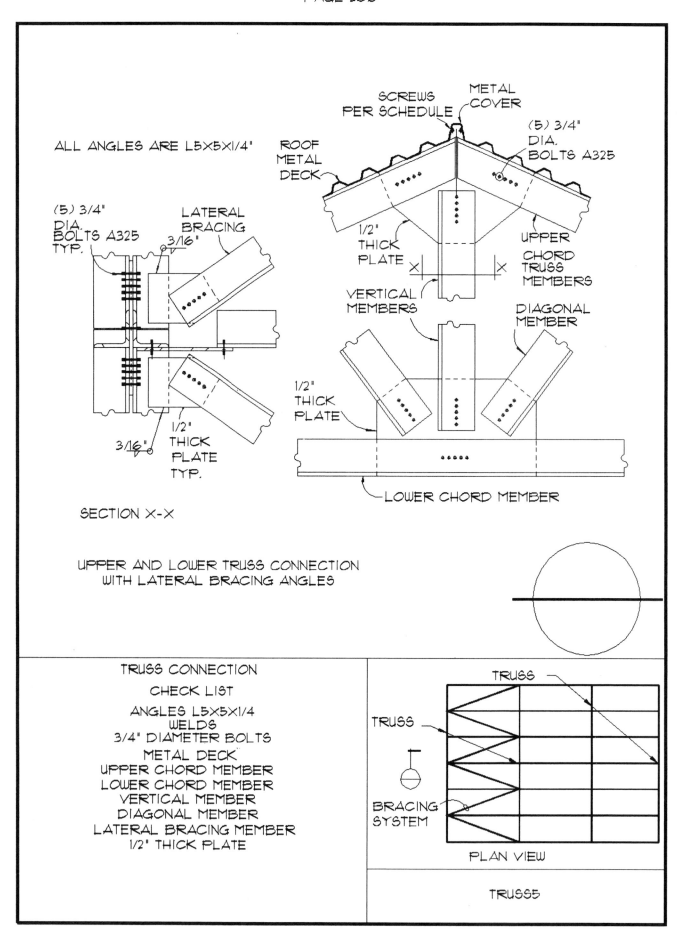

ALL ANGLES ARE L5X5X1/4"

(5) 3/4"
DIA.
BOLTS A325
TYP.

LATERAL
BRACING

3/16"

3/16"

1/2"
THICK
PLATE
TYP.

SCREWS
PER SCHEDULE

METAL
COVER

ROOF
METAL
DECK

(5) 3/4"
DIA.
BOLTS A325

1/2"
THICK
PLATE

UPPER
CHORD
TRUSS
MEMBERS

VERTICAL
MEMBERS

DIAGONAL
MEMBER

1/2"
THICK
PLATE

LOWER CHORD MEMBER

SECTION X-X

UPPER AND LOWER TRUSS CONNECTION
WITH LATERAL BRACING ANGLES

TRUSS CONNECTION
CHECK LIST
ANGLES L5X5X1/4
WELDS
3/4" DIAMETER BOLTS
METAL DECK
UPPER CHORD MEMBER
LOWER CHORD MEMBER
VERTICAL MEMBER
DIAGONAL MEMBER
LATERAL BRACING MEMBER
1/2" THICK PLATE

TRUSS

TRUSS

BRACING
SYSTEM

PLAN VIEW

TRUSS5

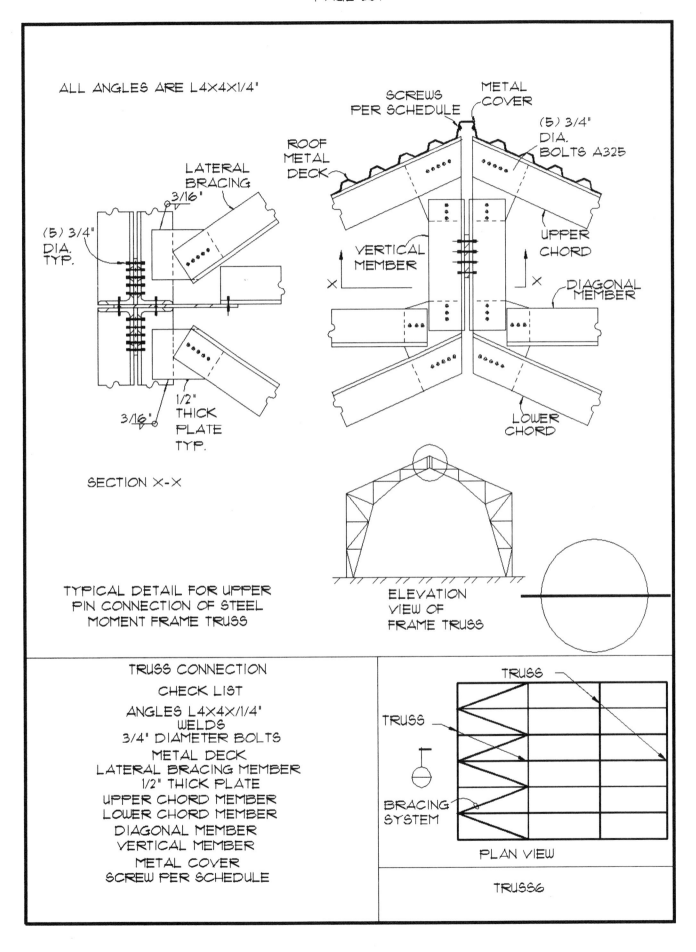

ALL ANGLES ARE L4X4X1/4"

LATERAL
BRACING

3/16"

(5) 3/4"
DIA.
TYP.

3/16"

1/2"
THICK
PLATE
TYP.

SECTION X-X

TYPICAL DETAIL FOR UPPER
PIN CONNECTION OF STEEL
MOMENT FRAME TRUSS

SCREWS
PER SCHEDULE

METAL
COVER

ROOF
METAL
DECK

(5) 3/4"
DIA.
BOLTS A325

VERTICAL
MEMBER

X

X

UPPER
CHORD

DIAGONAL
MEMBER

LOWER
CHORD

ELEVATION
VIEW OF
FRAME TRUSS

TRUSS CONNECTION
CHECK LIST
ANGLES L4X4X/1/4"
WELDS
3/4" DIAMETER BOLTS
METAL DECK
LATERAL BRACING MEMBER
1/2" THICK PLATE
UPPER CHORD MEMBER
LOWER CHORD MEMBER
DIAGONAL MEMBER
VERTICAL MEMBER
METAL COVER
SCREW PER SCHEDULE

TRUSS

TRUSS

BRACING
SYSTEM

PLAN VIEW

TRUSS6

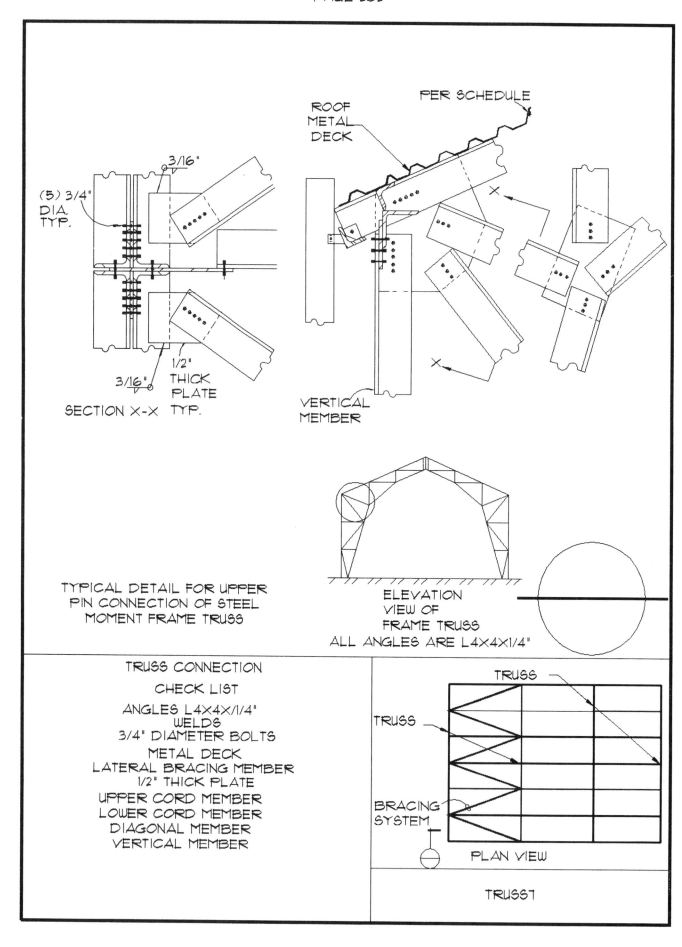

3/16"

(5) 3/4"
DIA.
TYP.

3/16"

SECTION X-X

1/2"
THICK
PLATE
TYP.

ROOF
METAL
DECK

PER SCHEDULE

VERTICAL
MEMBER

X

X

TYPICAL DETAIL FOR UPPER
PIN CONNECTION OF STEEL
MOMENT FRAME TRUSS

ELEVATION
VIEW OF
FRAME TRUSS
ALL ANGLES ARE L4X4X1/4"

TRUSS CONNECTION

CHECK LIST

ANGLES L4X4X/1/4"
WELDS
3/4" DIAMETER BOLTS
METAL DECK
LATERAL BRACING MEMBER
1/2" THICK PLATE
UPPER CORD MEMBER
LOWER CORD MEMBER
DIAGONAL MEMBER
VERTICAL MEMBER

TRUSS

TRUSS

BRACING
SYSTEM

PLAN VIEW

TRUSST

PRODUCTS FROM CASA UTILITY

CASA Utility provides computer software for design and analysis. Here is a list of this software:

1 - RETAINING WALL SOFTWARE

This software is used to design and engineer retaining walls
with the following capabilities:

a - restrained or cantilevered retaining wall
b - concrete or masonry
c - inspected or non-inspected
d - option to use soil report data or Coulumb's equation
    to determine the soil pressure
e - option to select the reinforcement and have the program
    provide the section required
f - design of wall and footing
g - check for overturning and sliding
h - option to select the steel rebar
i - option to read and save files

2 - WOOD BEAM DESIGN SOFTWARE

This software is used to design or check the sizes of wood
beam members, with the following capabilities:

a - design or check the size of wood beams
b - cantilever or overhang
c - option to select lumber section or GLB
d - flat or taper section
e - option to limit the deflection to l/240 or l/360, or to
    any user-supplied limit
f - design according to UBC
g - option to input and save the material properties in
    separate files
h - option to input several point or linear loads
i - option for load combinations
j - option to factor the load
k - gives the user many options for each case of loading
l - output the size of section, bending stress, deflection,
    and shear stress
m - complete lumber and GLB size library
n - option to read and save files

3 - TILTUP AND SLENDER WALL SOFTWARE

This software is used to design and engineer slender walls
for tiltup concrete, masonry of brick walls according to the
Uniform Building Code, with the following capabilities:

a - option to change the UBC parameters and update the
    software according to current code
b - masonry, concrete, or brick
c - option to select the section of the rebar and the spacing
d - option to input the depth of the section
e - inspected or non-inspected
f - option to input axial point load with eccentricity
    or without eccentricity
g - option to input uniform load as wind, seismic, dead, or
    live load
h - load combinations
i - factored load combinations
j - option to save the date for the material properties in
    separate files
k - steel rebar library
l - check the moment capacity for each load combination
m - check the service load deflection for each load
    combination
n - option to read and save files

To get the demo disk for all above software, send $2.50 to
CASA Utility at:
        10307 Azuaga Street, Suite 32
        San Diego CA, 92129
        Tel: (619) 538-2376

Prices: One Software Package $199.00
         Two Software Packages $249.00
         Three Software Packages $299.00

INDEX

column	SBASE1-6, 9, 11-13, BRACKET1-6, BRACE1-7, 9, 11, 12, COMPO5, 6, DECK1, FND7, MOMENT1-20, OPENING2-3, PIN1, 4-7, ROOF1, 4-6, 12-19, SECTION1-5, 9, SHEAR1, 2, 5, 6, 8, 10, 11, 13, 14, 16, SLIDE1-11, SPLICE9-13, STAIR3, 7, 11,-16
concrete	SBASE1-13, BRACE1, 6, 12, COMPO5, 8, 9, 10-13, 14-16, DECK11, FND1-7, PIN7, SECTION6, 7, STAIR1, 5,8, 13-16, STUD1-6, TRUSS2
concrete slab	COMPO7, 10, 11, 12
concrete topping	COMPO2-1, 18, DECK2-4, 8-11, STAIR4, 13, 16, TRUSS2
connector	BRACE12
cylinder	SBASE6, 7, 8, PIN1-4, 6, TENSION5
diagonal angle	COMPO12
diagonal member	SBASE7, 8, BRACE1-13, COMPO13, DECK1, MOMENT18, 20, PIN1-5, ROOF10, SHEAR5, 8, 14, TENSION4, 5, TRUSS1, 2, 4-7
double angle	SBASE7, 8, BRACE3-10, TRUSS1-7
drag angle	COMPO12
edge nailing	MOMENT16
end member	SBASE7, 8
existing beam	MOMENT16
expansion connector	STUD4, TRUSS2, 4
fillet weld	SBASE1-13, BRACE1-13, BRACKET1-6, COMPO6, 11, 12-16, DECK2-7, MOMENT1-15, 17-21, OPENING1-3, 5, PIN1-6, 7, ROOF1-24, SECTION2, 4, 6, 8, 9, SHEAR1-9, 11-13, 16, SLIDE1-9, 11, SPLICE2, 3, 6, 7, 9, 12, 13, 15, 16, STAIR1-17, TENSION1-5, TORSION1-3, TRUSS1-7

fill	OPENING2, SPLICE9-11, 14
footing	SBASE9, FND7
footing reinforcement	FND7
full penetration weld	SBASE12, BRACE2, 9, 11, MOMENT1, 4, 6-10, 16-18, 20, SHEAR6, 7, 10, SLIDE11, SPLICE6-8, 12, 13, 15, 17, STAIR2, 9, 11, 14, TORSION1
gap	COMPO16, PIN6, SLIDE1-11, STUD5-6
grade beam	FND1-7
grout	SBASE13, STAIR14
handrail	STAIR10, 17
horizontal reinforcement	COMPO13, 14, ROOF24, SECTION6, 7, SLIDE6
horizontal stirrups	FND1-13
lateral bracing	BARCE1-13
ledger angle	COMPO13
lower cord member	SBASE7, 8
main slab reinforcement	COMPO8
metal connector with vertical movement	DECK11
metal connector	COMPO15, ROOF1, 8-11, 24
metal deck	COMPO2-5, 13-16, DECK1-11, MOMENT20, OPENING6, 7, ROOF1-11, 14-24, STAIR1-4, 6-9, 12-17, TRUSS4-7
metal studs	DECK10, 11, MOMENT19, 20, ROOF1, 8, 11, 20, 24, STUD1-11, TRUSS2-7
mortar	SBASE5
nut	TENSION1, 2

shear resisting angle	COMPO13, 14, 16, DECK2, 4-7, 11, ROOF1, 3, 5, 6, 8, 11, 22, 23, STAIR4
shot pin	DECK10, STUD4-6
shrinkage reinforcement	COMPO3, 4, 7, FND1-6
slab	COMPO7-12, DECK10, ROOF24, SECTION6-7, SLIDE6, STAIR4, 5, 8, 14-16, STUD5
slotted holes	MOMENT7, SLIDE1, 2, 6-11
splice	FND6
stiffener	SBASE6, 7, 8, 11, 12, 13, BRACE2, 3, 4, 5, 9, 10, 13, BRACKET1-6, COMPO1, 13, MOMENT1-3, 5-15, 17, 18, 20, OPENING1-3, PIN, 4, ROOF4, 6, 10, 11, 13, 20, SECTION6-9, SLIDE6, SPLICE3, STAIR4, 5, 17, TRUSS1, 2, 3
stirrups	SBASE9, COMPO3-7, 10, 11, FND1, 4, 6, 7, STAIR8
stop pour	DECK2
strap	STUD7, 9
t section	MOMENT4, 14, 15, SHEAR16, SLIDE10
threaded connection	PIN2-4
top base plate	SBASE10
top beam reinforcement	COMPO3, 4, 10, 11, STAIR8
top chord angles	COMPO13, TRUSS1, 2, 4-7
top chord truss	COMPO13, TRUSS1, 2, 4-7
top flange plate	BRACE2, 11, BRACKET4, MOMENT1-4, 9, 10, 17, 18, SECTION6-9, SPLICE1-3, 5, 6
top steel reinforcement	COMPO2-4, 7, 10, 11, DECK2, 3, 8-11, FND1-7, TRUSS2